Introduction to Elasticity, Fracture and Fatigue

Kang Yong Lee

# Introduction to Elasticity, Fracture and Fatigue

Springer

Kang Yong Lee
Seoul, Korea (Republic of)

ISBN 978-981-96-3602-0 ISBN 978-981-96-3600-6 (eBook)
https://doi.org/10.1007/978-981-96-3600-6



This Springer imprint is published by the registered company Springer Nature Singapore Pte Ltd.
The registered company address is: 152 Beach Road, #21-01/04 Gateway East, Singapore 189721, Singapore

# Preface

As indicated by its title, *Introduction to Elasticity, Fracture, and Fatigue*, this book covers three major topics, elasticity, fracture, and fatigue. The content is adapted from my lecture notes, originally developed for third- and fourth-year undergraduate students. A foundational knowledge of Mechanics of Materials is required to understand the topics covered in this book. This book is divided into three parts, elasticity, fracture, and fatigue. Part I deals with elasticity and consists of four chapters. Chapter 1 introduces tensor notation. To familiarize students with tensors, topics like coordinate transformation, stress transformation, the Kronecker delta, and principal stress are presented in both tensor and expanded forms. In Chap. 2, three-dimensional governing equations in elasticity are introduced in tensor notation. This chapter includes the equilibrium equation, Hooke's law, the relationship between strain and displacement, traction vectors, compatibility equations, and Navier's equilibrium equations. Chapter 3 deals with two-dimensional elasticity. The difference of plane strain state and plane stress state is stated, and the governing equations of each state are obtained from three-dimensional governing equations. In Chap. 4, Airy stress functions in Cartesian and polar coordinates are introduced. The transformation method between the governing equations in Cartesian coordinate and polar coordinate is also explained. Part II treats fracture and consists of three chapters. In Chap. 5, the difference between notch and crack, the concept, physical meaning, and importance of stress intensity factor, and theoretical and experimental plastic zone shapes and sizes at the crack tip are explained. In Chap. 6, the standard test method of plane strain fracture toughness is introduced. The fracture mechanics design method to prevent yielding and fracture is also explained. Chapter 7 deals with fatigue fracture. It briefly introduces the standard testing method for fatigue crack growth rate. The difference between Paris' law and Forman's law is explained. How to search why a crack occurred in the structure and how to make fatigue crack growth retard are also introduced. Part III deals with fatigue subject and has only one chapter, Chap. 8. It explains why fatigue data have scattering band, how to get constant hysteresis loop from random fatigue load, how to get fatigue life, how to get approximation of S-N curves in metallic materials.

Only the references necessary to understand the contents are listed for each part in Bibliography.

Seoul, Korea (Republic of) Kang Yong Lee

# Contents

# About the Author

**Kang Yong Lee** is an Emeritus Professor at Yonsei University (YU) in Seoul, South Korea. He has held notable positions, including President of the Korean Society of Mechanical Engineers (KSME), Fellow Emeritus of The Korean Academy of Science and Technology, and Emeritus Member of The National Academy of Engineering of Korea. Additionally, he has served as a Professor of Engineering Mechanics at Dalian University of Technology, Dalian, China, and as a Lecturer in Aerospace Engineering at San Diego State University, California, USA. He was also an Editor of the *Indian Journal of Engineering and Materials Science.*

The author is listed in *Marquis Who's Who in the World* and was recognized by the *Outstanding People of the 20th Century* by the International Biographical Center, England, in 2000. He earned his BS and MS in Mechanical Engineering from YU in 1969 and 1971, respectively, and his Ph.D. in Mechanical Engineering from West Virginia University, USA, in 1978. Following his doctorate, he worked for two years at the Korean Institute of Machinery and Metals and for 32 years at YU.

He received the Yonsei Academic Awards from YU in 1987 and 2009, an Academic Award from KSME in 1988, and the Best Research Performance Professor Awards from YU in 1996, 2002–2005, 2007, 2008, and 2011. His research interests include fracture mechanics, fatigue, stress analysis, acoustic emission testing, microstructure evaluation, fretting corrosion, and welding. He has published over 300 SCI papers and authored four books, including *Multiple Crack Problems in Elasticity* (WIT Press, 2003), *Fracture and Strength '90* (TTP, 1991), *Design Against Fracture and Fatigue* (TTP, 2010), *Stress Analysis and Mechanical Behaviors of Materials* (YU Press, 2003, awarded by Korea Ministry of Culture and Tourism), and *Applied Mechanics of Materials* (YU Press, 2024).

# Part I
# Elasticity

# Chapter 1
# Tensor

**Abstract** Tensor notation is introduced. To familiarize students with tensors, topics like coordinate transformation, stress transformation, and principal stress are presented in both tensor and expanded forms.

## 1.1 Cartesian Tensor

Tensor notation is a mathematical tool that expresses equations concisely. In this book, only Cartesian tensors are used. Table 1.1 lists different tensor types.

As shown in Table 1.1, a zero order tensor has one component. It is called scalar. Volume, temperature, mass, etc. are scalar. A first-order tensor has three components and is called a vector. Force, velocity, moment, etc. are vector. For example, a force vector $\overline{F}$ is expressed as follows,

$$\overline{F} = F_x\vec{i} + F_y\vec{j} + F_z\vec{k} \tag{1.1}$$

where $F_x$, $F_y$, and $F_z$ are force components of $x$, $y$, and $z$ directions, respectively, and $\vec{i}, \vec{j}$ and $\vec{k}$ are the corresponding unit vectors.

A second-order tensor has nine components. Quantities such as stress and strain are examples of second-order tensors. For example, the stress tensor $\sigma_{\mathrm{ij}}$ is expressed as follows,

$$\begin{pmatrix} \sigma_{\mathrm{xx}} & \sigma_{\mathrm{xy}} & \sigma_{\mathrm{xz}} \\ \sigma_{\mathrm{yx}} & \sigma_{\mathrm{yy}} & \sigma_{\mathrm{yz}} \\ \sigma_{\mathrm{zx}} & \sigma_{\mathrm{zy}} & \sigma_{\mathrm{zz}} \end{pmatrix} \tag{1.2}$$

But in the Mechanics of Materials, the following form used to be expressed instead of Eq. 1.2,



K. Y. Lee, *Introduction to Elasticity, Fracture and Fatigue*,
https://doi.org/10.1007/978-981-96-3600-6_1

**Table. 1.1** Cartesian tensor

| Zero order tensor | Volume, temperature, mass, etc |
|---|---|
| = Scalar | (1 component) |
| First-order tensor | force, velocity, moment, etc |
| = Vector | (3 components) |
| Second-order tensor | stress, strain, etc |
| = Tensor | (9 components) |

$$\begin{pmatrix} \sigma_x & \tau_{\mathrm{xy}} & \tau_{\mathrm{xz}} \\ \tau_{\mathrm{yx}} & \sigma_y & \tau_{\mathrm{yz}} \\ \tau_{\mathrm{zx}} & \tau_{\mathrm{zy}} & \sigma_z \end{pmatrix} \tag{1.3}$$

However, in this book, the symbol of all the stresses regardless of normal and shear stresses is displayed as $\sigma$.

Higher-order tensors beyond the second order are also referred to as tensors. The number of components of an $n$-order tensor is calculated as $3^n$.

## 1.2 Coordinate Transformation

Coordinate transformation is represented by Eq. 1.4 in tensor notation.

$$x_j' = a_{\mathrm{ji}} x_i \quad \left(i = 1, 2, 3 \,\mathrm{and}\, j = 1', 2', 3'\right) \tag{1.4}$$

where $a_{\mathrm{ji}}$ is known as a direction cosine.

In Eq. 1.4, any index repeated twice within a term is referred to as a dummy index. Conversely, an index that does not repeat within a term is called a free index.

The direction cosine $a_{\mathrm{ji}}$ is defined as follows,

$$a_{\mathrm{ji}} = \cos(j, i), \tag{1.5}$$

where $(j, i)$ means the angle between positive $x_j\prime$ and $x_i$ axes.

Equation 1.4 indicates that a coordinate $x_i$ transforms to another coordinate $x_j\prime$.

In other word, the coordinate transformation gives the relationship between coordinates $x_i$ and $x_j\prime$.

When $j$ is replaced to $1'$, $2'$, and $3'$ in Eq. 1,4, it gives

$$\begin{cases} x_1\prime = a_{1'i} x_i \\ x_2\prime = a_{2'i} x_i \\ x_3\prime = a_{3'i} x_i \end{cases} \tag{1.6}$$

In Eq. 1.6, dummy index $i$ is treated by the following way,

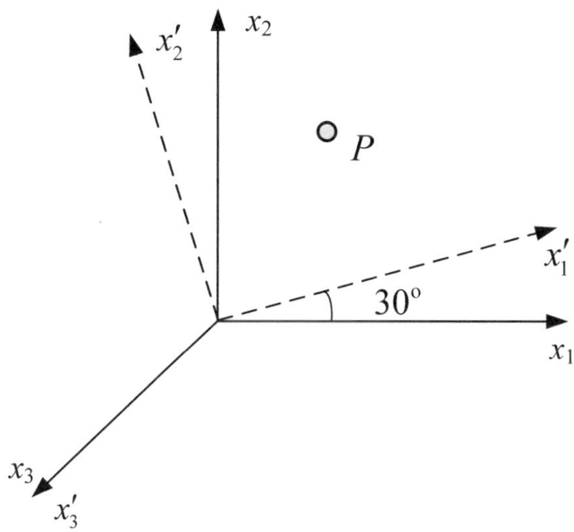


**Fig. 1.1** *P* point under two different coordinate systems

$$\begin{cases} x_{1'} = a_{1'i}x_i = a_{1'1}x_1 + a_{1'2}x_2 + a_{1'3}x_3 \\ x_{2'} = a_{2'i}x_i = a_{2'1}x_1 + a_{2'2}x_2 + a_{2'3}x_3 \\ x_{3'} = a_{3'i}x_i = a_{3'1}x_1 + a_{3'2}x_2 + a_{3'3}x_3 \end{cases} \quad (1.7)$$

This equation is known as an unabridged form of Eq. 1.4.

**Example 1.1** In Fig. 1.1, the point of *P* lies within the body. The coordinates of *P* depend on the coordinate system.

If the coordinates of $P(x_1, x_2, x_3)$ are (1,1,1), determine the coordinates of $P(x_{1'}, x_{2'}, x_{3'})$?

**Solution**

Apply Eq. 1.4, $x_{j'} = a_{\mathrm{ji}}x_i$.

$$\begin{aligned} x_{1'} &= a_{1'i}x_i = a_{1'1}x_1 + a_{1'2}x_2 + a_{1'3}x_3 \\ &= \cos(1', 1)x_1 + \cos(1', 2)x_2 + \cos(1', 3)x_3 \\ &= x_1 \cos 30^o + x_2 \cos 60^o + x_3 \cos 90^o \\ &= 1 \times \frac{\sqrt{3}}{2} + 1 \times \frac{1}{2} + 1 \times 0 \\ &= \frac{1 + \sqrt{3}}{2} \end{aligned}$$

$$\begin{aligned} x_{2'} &= a_{2'i}x_i = a_{2'1}x_1 + a_{2'2}x_2 + a_{2'3}x_3 \\ &= \cos(2', 1)x_1 + \cos(2', 2)x_2 + \cos(2', 3)x_3 \end{aligned}$$

$$
\begin{aligned}
&= x_1 \cos 120^o + x_2 \cos 30^o + x_3 \cos 90^o \\
&= 1 \times \left(-\frac{1}{2}\right) + 1 \times \frac{\sqrt{3}}{2} + 1 \times 0 \\
&= \frac{-1+\sqrt{3}}{2}
\end{aligned}
$$

$$
\begin{aligned}
x_3\prime &= a_{3'i}x_i = a_{3'1}x_1 + a_{3'2}x_2 + a_{3'3}x_3 \\
&= \cos(3', 1)x_1 + \cos(3', 2)x_2 + \cos(3', 3)x_3 \\
&= x_1 \cos 90^o + x_2 \cos 90^o + x_3 \cos 0^o \\
&= 1 \times (0) + 1 \times 0 + 1 \times 1 \\
&= 1
\end{aligned}
$$

Therefore, the coordinate of $P(x_1\prime, x_2\prime, x_3\prime)$ is $\left(\frac{1+\sqrt{3}}{2}, \frac{-1+\sqrt{3}}{2}, 1\right)$

## 1.3 Stress Transformation

Stress transformation is represented as follows,

$$
\sigma_{\alpha\beta}\prime = a_{\alpha i}a_{\beta j}\sigma_{\text{ij}} \quad \left(\alpha, \beta = x', y', z' \text{ and } i, j = x, y, z\right) \tag{1.8}
$$

where $a_{\alpha i}$ and $a_{\beta j}$ are direction cosines.

This defines the relationship between $\sigma_{ij}$ and $\sigma_{\alpha\beta}\prime$ at an arbitrary point within a body under two distinct coordinate systems, $x_i$ and $x_j\prime$.

In other word, $\sigma_{ij}$ in the $x_i$ coordinate system is transformed to $\sigma_{\alpha\beta}\prime$ in the $x_j\prime$ coordinate system.

The expanded forms of Eq. 1.8 can be derived using the method from Sect. 1.2, as follows,

$$
\begin{aligned}
\sigma_x' &= a_{x'x}^2\sigma_x + a_{x'y}^2\sigma_y + a_{x'z}^2\sigma_z + 2\left(a_{x'x}a_{x'y}\sigma_{xy} + a_{x'y}a_{x'z}\sigma_{yz} + a_{x'z}a_{x'x}\sigma_{zx}\right) \\
\sigma_y' &= a_{y'x}^2\sigma_x + a_{y'y}^2\sigma_y + a_{y'z}^2\sigma_z + 2\left(a_{y'x}a_{y'y}\sigma_{xy} + a_{y'y}a_{y'z}\sigma_{yz} + a_{y'z}a_{y'x}\sigma_{zx}\right) \\
\sigma_z' &= a_{z'x}^2\sigma_x + a_{z'y}^2\sigma_y + a_{z'z}^2\sigma_z + 2\left(a_{z'x}a_{z'y}\sigma_{xy} + a_{z'y}a_{z'z}\sigma_{yz} + a_{z'z}a_{z'x}\sigma_{zx}\right) \\
\sigma_{xy}' &= \sigma_{yx}' = a_{x'x}a_{y'x}\sigma_x + a_{x'y}a_{y'y}\sigma_y + a_{x'z}a_{y'z}\sigma_z + (a_{x'x}a_{y'y} + a_{y'x}a_{x'y})\sigma_{xy} \\
&\quad + (a_{x'y}a_{y'z} + a_{y'y}a_{x'z})\sigma_{yz} + (a_{x'z}a_{y'x} + a_{y'z}a_{x'x})\sigma_{zx} \\
\sigma_{yz}' &= \sigma_{zy}' = a_{y'x}a_{z'x}\sigma_x + a_{y'y}a_{z'y}\sigma_y + a_{y'z}a_{z'z}\sigma_z + \left(a_{y'x}a_{z'y} + a_{z'x}a_{y'y}\right)\sigma_{xy} \\
&\quad + \left(a_{y'y}a_{z'z} + a_{z'y}a_{y'z}\right)\sigma_{yz} + \left(a_{y'z}a_{z'x} + a_{z'z}a_{y'x}\right)\sigma_{zx}
\end{aligned}
$$

$$\sigma'_{zx} = \sigma'_{xz} = a_{z'x}a_{x'x}\sigma_x + a_{z'y}a_{x'y}\sigma_y + a_{z'z}a_{x'z}\sigma_z + \left(a_{z'x}a_{x'y} + a_{x'x}a_{z'y}\right)\sigma_{xy}$$
$$+ \left(a_{z'y}a_{x'z} + a_{x'y}a_{z'z}\right)\sigma_{yz} + \left(a_{z'z}a_{x'x} + a_{x'z}a_{z'x}\right)\sigma_{zx} \tag{1.9}$$

**Example 1.2** Using Eq. 1.8, find $\sigma_{x'}$ shown in Eq. 1.9.

**Solution**

$$\begin{aligned}
&\alpha, \beta = x' \\
&\sigma'_{x'x'} \equiv \sigma'_{xx} = a_{x'i}a_{x'j}\sigma_{ij} \\
&\quad = a_{x'x}a_{x'j}\sigma_{xj} + a_{x'y}a_{x'j}\sigma_{yj} + a_{x'z}a_{x'j}\sigma_{zj} \\
&\quad = a_{x'x}a_{x'x}\sigma_{xx} + a_{x'x}a_{x'y}\sigma_{xy} + a_{x'x}a_{x'z}\sigma_{xz} \\
&\quad + a_{x'y}a_{x'x}\sigma_{yx} + a_{x'y}a_{x'y}\sigma_{yy} + a_{x'y}a_{x'z}\sigma_{yz} \\
&\quad + a_{x'z}a_{x'x}\sigma_{zx} + a_{x'z}a_{x'y}\sigma_{zy} + a_{x'z}a_{x'z}\sigma_{zz} \\
&\quad = a^2_{x'x}\sigma_{xx} + a^2_{x'y}\sigma_{yy} + a^2_{x'z}\sigma_{zz} + 2\left(a_{x'x}a_{x'y}\sigma_{xy} + a_{x'x}a_{x'z}\sigma_{xz} + a_{x'y}a_{x'z}\sigma_{yz}\right)
\end{aligned}$$

$$\therefore\ \sigma_{x'} = a^2_{x'x}\sigma_x + a^2_{x'y}\sigma_y + a^2_{x'z}\sigma_z + 2\left(a_{x'x}a_{x'y}\sigma_{xy} + a_{x'x}a_{x'z}\sigma_{xz} + a_{x'y}a_{x'z}\sigma_{yz}\right)$$

This equation is derived by substituting $\sigma_{xx'}$ with $\sigma_{x'}$, $\sigma_{xx}$ with $\sigma_x$, $\sigma_{yy}$ with $\sigma_y$, and $\sigma_{zz}$ with $\sigma_z$, applying the following relationship

$$\sigma_{xy} = \sigma_{yx},\ \sigma_{xz} = \sigma_{zx},\ \sigma_{yz} = \sigma_{zy} \tag{1.10}$$

Equation 1.10 can be verified through moment equilibrium consideration.

## 1.4 Kronecker Delta

Kronecker delta is defined as follows,

$$\delta_{ij} = \begin{cases} 1, & i = j \\ 0, & i \neq j \end{cases} \quad (i, j = 1, 2, 3) \tag{1.11}$$

In other word,

$$\begin{cases} \delta_{11}, \delta_{22}, \delta_{33} = 1 \\ \delta_{12}, \delta_{21}, \delta_{13}, \delta_{31}, \delta_{23}, \delta_{32} = 0 \end{cases} \tag{1.12}$$

When expressing the transformation from $x_k$ to $x_i$ is required, Kronecker delta is used as follows,

$$x_i = \delta_{ik}x_k \tag{1.13}$$

**Example 1.3** Determine the value of $\delta_{ii}$?

**Solution**

Since $i$ is a dummy index, we can write

$$\delta_{ii} = \delta_{11} + \delta_{22} + \delta_{33} = 3$$

**Example 1.4** Verify that Eq. 1.13 is correct.

**Solution**

We can verify the accuracy of Eq. 1.13 as follows,

$$x_i = \delta_{ik} x_k = \delta_{i1} x_1 + \delta_{i2} x_2 + \delta_{i3} x_3 \quad (i, k = 1, 2, 3)$$

$$\begin{cases} x_1 = \delta_{11} x_1 + \delta_{12} x_2 + \delta_{13} x_3 = x_1 + 0 + 0 = x_1 \\ x_2 = \delta_{21} x_1 + \delta_{22} x_2 + \delta_{23} x_3 = 0 + x_2 + 0 = x_2 \\ x_3 = \delta_{31} x_1 + \delta_{32} x_2 + \delta_{33} x_3 = 0 + 0 + x_3 = x_3 \end{cases}$$

## 1.5 Principal Stress

Principal stress is determined using the following equation,

$$\left|\sigma_{\mathrm{ij}} - S\delta_{\mathrm{ij}}\right| = 0 \quad (i, j = x, y, z) \tag{1.14}$$

where $\sigma_{\mathrm{ij}}$ represents stress components at an arbitrary point $P$ within the body under applied load, $S$ is principal stress at the point $P$ and $\delta_{\mathrm{ij}}$ is Kronecker delta.

The expanded form of Eq. 1.14 is expressed as follows,

$$\begin{vmatrix} (\sigma_x - S) & \sigma_{\mathrm{xy}} & \sigma_{\mathrm{xz}} \\ \sigma_{\mathrm{yx}} & (\sigma_y - S) & \sigma_{\mathrm{yz}} \\ \sigma_{\mathrm{zx}} & \sigma_{\mathrm{zy}} & (\sigma_z - S) \end{vmatrix} = 0 \tag{1.15}$$

The determinant is calculated as follows,

$$\begin{vmatrix} a & b & c \\ d+ & e- & f \\ g & h & i \end{vmatrix} = (aei + bfg + cdh) - (ceg + afh + bdi) \tag{1.16}$$

By applying Eqs. 1.15 to 1.16, the following equation can be obtained,

$$S^3 - I_1 S^2 - I_2 S - I_3 = 0 \tag{1.17}$$

where

$$
\begin{aligned}
I_1 &= \sigma_x + \sigma_y + \sigma_z \\
I_2 &= -\left(\sigma_x\sigma_y + \sigma_y\sigma_z + \sigma_z\sigma_x\right) + \left(\sigma_{xy}^2 + \sigma_{yz}^2 + \sigma_{zx}^2\right) \\
I_3 &= \sigma_x\sigma_y\sigma_z + 2\sigma_{xy}\sigma_{yz}\sigma_{zx} - \left(\sigma_{yz}^2\sigma_x + \sigma_{zx}^2\sigma_y + \sigma_{xy}^2\sigma_z\right) \\
S &= \sigma_1, \sigma_2, \sigma_3 \quad (\sigma_1 > \sigma_2 > \sigma_3)
\end{aligned}
\tag{1.18}
$$

Principal stress is a type of normal stress that acts on a principal plane. A principal plane experiences only normal stress and no shear stress. If $\sigma_1, \sigma_2, \sigma_3$ are principal stresses, Eq. 1.15 can be written as,

$$
\begin{vmatrix}
(\sigma_1 - S) & 0 & 0 \\
0 & (\sigma_2 - S) & 0 \\
0 & 0 & (\sigma_3 - S)
\end{vmatrix} = 0
\tag{1.19}
$$

Equation 1.19 becomes $(\sigma_1 - S)(\sigma_2 - S)(\sigma_3 - S) = 0$. It gives Eq. 1.20.

$$
S^3 - I_1\prime S^2 - I_2\prime S - I_3\prime = 0
\tag{1.20}
$$

where

$$
\begin{aligned}
I_1\prime &= \sigma_1 + \sigma_2 + \sigma_3 \\
I_2\prime &= -(\sigma_1\sigma_2 + \sigma_2\sigma_3 + \sigma_3\sigma_1) \\
I_3\prime &= \sigma_1\sigma_2\sigma_3
\end{aligned}
\tag{1.21}
$$

Since $(\sigma_1, \sigma_2, \sigma_3)$ comes from $\sigma_{ij}$,$(\sigma_1, \sigma_2, \sigma_3)$ is equivalent to $\sigma_{ij}$.
In other word, Eq. 1.20 should be the same as Eq. 1.17.
Therefore,

$$
I_1 = I_1\prime, I_2 = I_2\prime, I_3 = I_3\prime
\tag{1.22}
$$

$I_1$, $I_{2,}$ and $I_3$ in Eq. 1.18 arc expressed with stress components $\sigma_{ij}$ under the $(x, y, z)$ coordinate system. But $I_1\prime, I_{2,}\prime\, and\, I_3\prime$ in Eq. 1.21 are expressed with principal stresses under the principal axes $(1, 2, 3)$ coordinate system.

Equation 1.22 means that the quantities of $I_1$, $I_2$ and $I_3$ cannot change although coordinate system changes from $(x, y, z)$ to $(1, 2, 3)$ coordinate system. Therefore, these values are referred to as stress invariants because, despite changes in the coordinate system, they remain constant.

If ordinary stresses, $\sigma_{\text{ij}}$, instead of principal stresses, are given as a solution, the used coordinate system should be specified. However, if principal stresses are given as a result, it is not necessary to specify the used coordinate system, because the principal coordinate system is unique.

It is therefore recommended to present principal stresses $\sigma_1$, $\sigma_2$ and $\sigma_3$ rather than the ordinary stress components.

It is important to verify that the principal stresses derived from Eq. 1.17 satisfy Eq. 1.22.

## 1.6 Effective Stress

The effective stress, $\sigma_{\text{eff}}$, is sometimes referred to as equivalent stress or von Mises stress. It is defined as follows,

$$\sigma_{\text{eff}} = \frac{1}{\sqrt{2}}\left\{(\sigma_1 - \sigma_2)^2 + (\sigma_2 - \sigma_3)^2 + (\sigma_3 - \sigma_1)^2\right\}^{1/2} \tag{1.23}$$

where $\sigma_1$, $\sigma_2$ and $\sigma_3$ are principal stresses.

This equation is used to identify the yielding zone within the body under applied loads. If $\sigma_{\text{eff}}$ at a point in the body is greater than the material's yield stress, the point is considered yielded. If $\sigma_{\text{eff}}$ at the point is less than the yield stress, the point is not yielded. If $\sigma_{\text{eff}}$ equals the yield stress, the point is just at the yield limit. Using this approach, we can determine the plastic zone, elastic zone, and the boundary between these zones, as shown in Fig. 1.2.

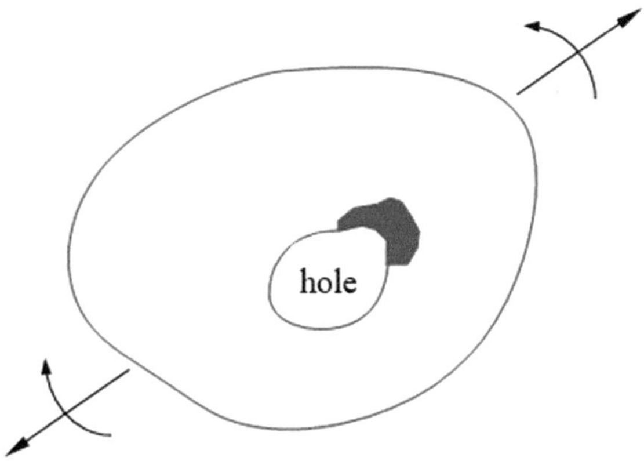


**Fig. 1.2** Yield zone represented by the shaded area

# Chapter 2
# Three-Dimensional Governing Equations

**Abstract** All kinds of three dimensional governing equations in elasticity are expressed in tensor notations.

In the theory of elasticity, there are fifteen governing equations: six for stress components, six for strain components, and three for displacement components. To determine these fifteen components, fifteen equations are required. As a result, three-dimensional elasticity problems are complex and challenging to solve.

## 2.1 Equilibrium Equations

The equilibrium equations are expressed as follows,

$$\sigma_{\mathrm{ij},j} + X_i = 0 \tag{2.1}$$

where $\sigma_{\mathrm{ij},j} \equiv \frac{\partial \sigma_{\mathrm{ij}}}{\partial x_j}$ and $X_i$ is body force per unit volume.

The unabridged form of Eq. 2.1 is written as follows,

$$\begin{aligned} \frac{\partial \sigma_x}{\partial x} + \frac{\partial \sigma_{\mathrm{xy}}}{\partial y} + \frac{\partial \sigma_{\mathrm{xz}}}{\partial z} + X = 0 \\ \frac{\partial \sigma_{\mathrm{yx}}}{\partial x} + \frac{\partial \sigma_y}{\partial y} + \frac{\partial \sigma_{\mathrm{yz}}}{\partial z} + Y = 0 \\ \frac{\partial \sigma_{\mathrm{zx}}}{\partial x} + \frac{\partial \sigma_{\mathrm{zy}}}{\partial y} + \frac{\partial \sigma_z}{\partial z} + Z = 0 \end{aligned} \tag{2.2}$$

Equation 2.2 represents the force equilibrium equation. The moment equilibrium equation is expressed as follows,

$$\sigma_{\mathrm{xy}} = \sigma_{\mathrm{yx}}, \sigma_{\mathrm{xz}} = \sigma_{\mathrm{zx}}, \sigma_{\mathrm{yz}} = \sigma_{\mathrm{zy}} \tag{2.3}$$

 
K. Y. Lee, *Introduction to Elasticity, Fracture and Fatigue*,
https://doi.org/10.1007/978-981-96-3600-6_2

If Eqs. 2.2 and 2.3 are satisfied, the body remains stationary, with no translational or rotational motion. Substituting Eq. 2.3 into Eq. 2.2 results in the following,

$$\begin{aligned}
&\frac{\partial \sigma_x}{\partial x}+\frac{\partial \sigma_{\mathrm{xy}}}{\partial y}+\frac{\partial \sigma_{\mathrm{xz}}}{\partial z}+X=0 \\
&\frac{\partial \sigma_{\mathrm{xy}}}{\partial x}+\frac{\partial \sigma_y}{\partial y}+\frac{\partial \sigma_{\mathrm{yz}}}{\partial z}+Y=0 \\
&\frac{\partial \sigma_{\mathrm{xz}}}{\partial x}+\frac{\partial \sigma_{\mathrm{yz}}}{\partial y}+\frac{\partial \sigma_z}{\partial z}+Z=0
\end{aligned} \tag{2.4}$$

We refer to Eq. 2.4 as the equilibrium equations.

## 2.2 Hooke's Law

Hooke's law defines the relationship between six stress components and six strain components. In linear elasticity, it is expressed as follows,

$$\sigma_{\mathrm{ij}}=\lambda \delta_{\mathrm{ij}} \varepsilon_{\alpha \alpha}+2 G \varepsilon_{\mathrm{ij}} \tag{2.5}$$

where $\lambda$ and $G$ are Lamé constants, with $G$ typically referred to as the shear modulus, and $\delta_{ij}$ is Kronecker delta.

An alternative form of Eq. 2.5 can be derived as follows,

$$\varepsilon_{\mathrm{ij}}=\frac{1+v}{E} \sigma_{\mathrm{ij}}-\frac{v}{E} \delta_{ij} \sigma_{\alpha \alpha}. \tag{2.6}$$

where $E$ is Young's modulus and $v$ is Poisson ratio.

The relationships among elastic constants are presented in Eq. 2.7.

$$\begin{aligned}
\lambda &= \frac{2 G v}{1-2 v}=\frac{G(E-2 G)}{3 G-E}=K-\frac{2}{3} G=\frac{E v}{(1+v)(1-2 v)}=\frac{3 K v}{1+v} \\
&= \frac{3 K(3 K-E)}{9 K-E} \\
G &= \frac{\lambda(1-2 v)}{2 v}=\frac{3}{2}(K-\lambda)=\frac{E}{2(1+v)}=\frac{3 K(1-2 v)}{2(1+v)}=\frac{3 \mathrm{KE}}{9 K-E} \\
v &= \frac{\lambda}{2(\lambda+G)}=\frac{\lambda}{(3 K-\lambda)}=\frac{E}{2 G}-1=\frac{3 K-2 G}{2(3 K+G)}=\frac{3 K-E}{6 K} \\
E &= \frac{G(3 \lambda+2 G)}{\lambda+G}=\frac{\lambda(1+v)(1-2 v)}{v}=\frac{9 K(K-\lambda)}{3 K-\lambda} \\
&= 2 G(1+v)=\frac{9 K G}{3 K+G}=3 K(1-2 v)
\end{aligned} \tag{2.7}$$

$$K = \lambda + \frac{2}{3}G = \frac{\lambda(1+v)}{3v} = \frac{3G(1+v)}{3(1-2v)} = \frac{GE}{3(3G-E)}$$
$$= \frac{E}{3(1-2v)}$$

where $K$ is bulk modulus.

The expanded forms of Eq. 2.6 are written as follows,

$$\begin{aligned}
\varepsilon_x &= \frac{1}{E}\left[\sigma_x - \nu\left(\sigma_y + \sigma_z\right)\right] \\
\varepsilon_y &= \frac{1}{E}\left[\sigma_y - \nu\left(\sigma_x + \sigma_z\right)\right] \\
\varepsilon_z &= \frac{1}{E}\left[\sigma_z - \nu\left(\sigma_x + \sigma_y\right)\right] \\
\varepsilon_{xy} &= \frac{\sigma_{xy}}{2G} \\
\varepsilon_{yz} &= \frac{\sigma_{yz}}{2G} \\
\varepsilon_{zx} &= \frac{\sigma_{zx}}{2G}
\end{aligned} \tag{2.8}$$

$\varepsilon_{xy}$, $\varepsilon_{yz}$, and $\varepsilon_{xz}$ in Eq. 2.8 is not real shear strain. The relationship between these strains and the real shear trains are shown as follows,

$$\varepsilon_{xy} = \frac{\gamma_{xy}}{2}, \quad \varepsilon_{yz} = \frac{\gamma_{yz}}{2}, \quad \varepsilon_{xz} = \frac{\gamma_{xz}}{2} \tag{2.9}$$

where $\gamma$ is a real shear strain.

Figure 2.1 explains $\gamma_{xy}$.

$\gamma_{xy}$ is an angle in radian. It represents the angle by which the plane defined by the x and y axes, as shown in Fig. 2.1, is distorted. The tensor notation of Eq. 2.9 is written as follows,

**Fig. 2.1** Shear strain

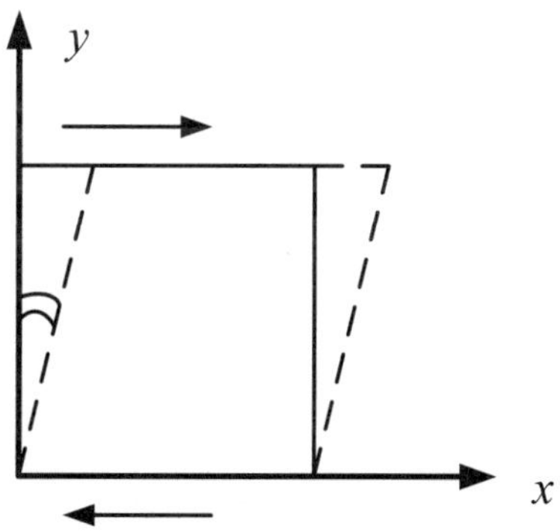

$$2\varepsilon_{\mathrm{ij}} = \gamma_{\mathrm{ij}}\,(i \neq j) \tag{2.10}$$

***Example 2.1*** Derive Eq. 2.6 from Eq. 2.5.

**Solution**

Equation 2.5 is modified as follows,

$$\varepsilon_{\mathrm{ij}} = \frac{1}{2G}\left(\sigma_{\mathrm{ij}} - \lambda\delta_{\mathrm{ij}}\varepsilon_{\alpha\alpha}\right)\,(a)$$

First, express $\varepsilon_{\alpha\alpha}$ in term of stress.

When the indexes $i$ and $j$ in Eq. 2.5 are replaced into $\alpha$, it becomes $\sigma_{\alpha\alpha} = \lambda\delta_{\alpha\alpha}\varepsilon_{\alpha\alpha} + 2G\varepsilon_{\alpha\alpha}$. However, it is incorrect because $\alpha$ is repeated four times within the term $\delta_{\alpha\alpha}\varepsilon_{\alpha\alpha}$. Therefore, it is re-written as follows, $\sigma_{\alpha\alpha} = \lambda\delta_{\beta\beta}\varepsilon_{\alpha\alpha} + 2G\varepsilon_{\alpha\alpha}$.

Since $\delta_{\beta\beta} = 3$, it is written as follows, $\sigma_{\alpha\alpha} = 3\lambda\varepsilon_{\alpha\alpha} + 2G\varepsilon_{\alpha\alpha}$.

Therefore, from this equation, $\varepsilon_{\alpha\alpha} = \frac{\sigma_{\alpha\alpha}}{3\lambda+2G}$. Introducing this equation into Equation a, we obtain $\varepsilon_{\mathrm{ij}} = \frac{1}{2G}\left(\sigma_{\mathrm{ij}} - \lambda\delta_{\mathrm{ij}}\frac{\sigma_{\alpha\alpha}}{3\lambda+2G}\right)$. By applying Eq. 2.7 to this result, we can get Eq. 2.6 finally.

## 2.3 Relationship Between Strain and Displacement

The relationship between strain and displacement is written in tensor form,

$$\varepsilon_{\mathrm{ij}} = \frac{1}{2}\left(u_{i,j} + u_{j,i}\right) \tag{2.11}$$

Equation 2.12 is the unabridged form of Eq. 2.11.

$$\begin{aligned}
\varepsilon_x &= \frac{\partial u_x}{\partial x}\\
\varepsilon_y &= \frac{\partial u_y}{\partial y}\\
\varepsilon_z &= \frac{\partial u_z}{\partial z}\\
\varepsilon_{\mathrm{xy}} &= \frac{1}{2}\left(\frac{\partial u_x}{\partial y} + \frac{\partial u_y}{\partial x}\right)\\
\varepsilon_{\mathrm{yz}} &= \frac{1}{2}\left(\frac{\partial u_y}{\partial z} + \frac{\partial u_z}{\partial y}\right)
\end{aligned} \tag{2.12}$$

So far, we have derived three equilibrium equations, six strain–stress relationships (known as Hooke's Law), and six strain–displacement relationships, totaling 15 equations.

***Example 2.2*** Given the displacement field equations, how can the strain and stress field equations be determined?

**Solution**

By substituting the given displacement field equations into Eq. 2.12, the strain field equations can be obtained. Then, by introducing these results into Eq. 2.5, the stress field equations are determined.

***Example 2.3*** Given the strain field equations, how can the stress and displacement field equations be determined?

**Solution**

By substituting the given strain field equations into Eq. 2.5, the stress field equations can be obtained. However, deriving the displacement field equations from the strain field equations requires integration, as shown in Eq. 2.12. This integration introduces constants that must be determined using boundary conditions. Applying integration along with boundary conditions yields the displacement field equations.

## 2.4 Traction Vector

Traction is a type of stress acting on a sectional area within the deformed body, as shown in Fig. 2.2. In this figure, $n$ represents an outward normal unit vector.

Traction vector is defined as follows,

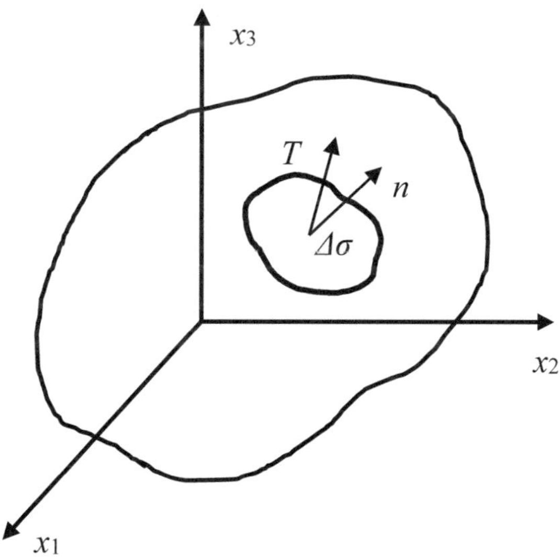


**Fig. 2.2** Traction vector

$$T_i = \sigma_{ij} n_j \tag{2.13}$$

where $n_j$ is defined as cos($n$,$j$), where ($n$,$j$) represents the angle between the positive $n$- and $j$- axes.

The unabridged form of Eq. 2.13 is as follows,

$$\begin{cases} T_x = \sigma_x n_x + \sigma_{xy} n_y + \sigma_{xz} n_z \\ T_y = \sigma_{yx} n_x + \sigma_y n_y + \sigma_{yz} n_z \\ T_z = \sigma_{zx} n_x + \sigma_{zy} n_y + \sigma_z n_z \end{cases} \tag{2.14}$$

## 2.5 Compatibility Equations

Compatibility ensures that deformation can physically occur in a deformable body under load. Compatibility equations are necessary to achieve a unique elastic solution. Although six strain field equations are provided, they may not inherently be compatible. For a unique, physically possible elastic solution, the strain field equations must relate to one another. These relationships are known as compatibility equations and are derived from Eq. 2.15.

$$\varepsilon_{\mathrm{ij}} = \frac{1}{2}\left(u_{i,j} + u_{j,i}\right) \tag{2.15}$$

From Eq. 2.15, we can get Eqs. 2.16.

$$\begin{aligned} \varepsilon_{\mathrm{kl,ij}} &= \frac{1}{2}\left(u_{k,\mathrm{lij}} + u_{l,\mathrm{kij}}\right) \\ \varepsilon_{\mathrm{kl,ij}} &= \frac{1}{2}\left(u_{k,\mathrm{lij}} + u_{l,\mathrm{kij}}\right) \\ \varepsilon_{\mathrm{jl,ik}} &= \frac{1}{2}\left(u_{j,\mathrm{lik}} + u_{l,\mathrm{jik}}\right) \\ \varepsilon_{\mathrm{ik,jl}} &= \frac{1}{2}\left(u_{i,\mathrm{kjl}} + u_{k,\mathrm{ijl}}\right) \end{aligned} \tag{2.16}$$

By adding and subtracting each other of Eq. 2.16, Eq. 2.17 can be obtained.

$$\varepsilon_{\mathrm{ij,kl}} + \varepsilon_{\mathrm{kl,ij}} - \varepsilon_{\mathrm{ik,jl}} - \varepsilon_{\mathrm{jl,ik}} = 0 \tag{2.17}$$

Equation 2.17 is called Saint–Venant compatibility equation.

As Eq. 2.17 is a fourth-order tensor, it comprises eighty-one equations. However, only Eq. 2.18 is independent among them.

$$\begin{cases} \frac{\partial^2 \varepsilon_x}{\partial y \partial z} = \frac{\partial}{\partial x}\left(-\frac{\partial \varepsilon_{yz}}{\partial x} + \frac{\partial \varepsilon_{zx}}{\partial y} + \frac{\partial \varepsilon_{xy}}{\partial z}\right) \\ \frac{\partial^2 \varepsilon_y}{\partial z \partial x} = \frac{\partial}{\partial y}\left(-\frac{\partial \varepsilon_{zx}}{\partial y} + \frac{\partial \varepsilon_{xy}}{\partial z} + \frac{\partial \varepsilon_{yz}}{\partial x}\right) \\ \frac{\partial^2 \varepsilon_z}{\partial x \partial y} = \frac{\partial}{\partial z}\left(-\frac{\partial \varepsilon_{xy}}{\partial z} + \frac{\partial \varepsilon_{yz}}{\partial x} + \frac{\partial \varepsilon_{zx}}{\partial y}\right) \\ 2\frac{\partial^2 \varepsilon_{xy}}{\partial x \partial y} = \frac{\partial^2 \varepsilon_x}{\partial y^2} + \frac{\partial^2 \varepsilon_y}{\partial x^2} \\ 2\frac{\partial^2 \varepsilon_{yz}}{\partial y \partial z} = \frac{\partial^2 \varepsilon_y}{\partial z^2} + \frac{\partial^2 \varepsilon_z}{\partial y^2} \\ 2\frac{\partial^2 \varepsilon_{xz}}{\partial x \partial z} = \frac{\partial^2 \varepsilon_z}{\partial x^2} + \frac{\partial^2 \varepsilon_x}{\partial z^2} \end{cases} \tag{2.18}$$

Through extensive manipulation of Eq. 2.18, we arrive at Eq. 2.19, known as the Beltrami-Michell compatibility equation."

$$\begin{cases} \nabla^2 \sigma_x + \frac{1}{1+\nu}\frac{\partial^2 \theta}{\partial x^2} = -\frac{\nu}{1-\nu}\mathrm{div}\mathbf{X} - 2\frac{\partial X_x}{\partial x} \\ \nabla^2 \sigma_y + \frac{1}{1+\nu}\frac{\partial^2 \theta}{\partial y^2} = -\frac{\nu}{1-\nu}\mathrm{div}\mathbf{X} - 2\frac{\partial X_y}{\partial y} \\ \nabla^2 \sigma_z + \frac{1}{1+\nu}\frac{\partial^2 \theta}{\partial z^2} = -\frac{\nu}{1-\nu}\mathrm{div}\mathbf{X} - 2\frac{\partial X_z}{\partial z} \\ \nabla^2 \sigma_{yz} + \frac{1}{1+\nu}\frac{\partial^2 \theta}{\partial y \partial z} = -\left(\frac{\partial X_y}{\partial z} + \frac{\partial X_z}{\partial y}\right) \\ \nabla^2 \sigma_{zx} + \frac{1}{1+\nu}\frac{\partial^2 \theta}{\partial z \partial x} = -\left(\frac{\partial X_z}{\partial x} + \frac{\partial X_x}{\partial z}\right) \\ \nabla^2 \sigma_{xy} + \frac{1}{1+\nu}\frac{\partial^2 \theta}{\partial x \partial y} = -\left(\frac{\partial X_x}{\partial y} + \frac{\partial X_y}{\partial x}\right) \end{cases} \tag{2.19}$$

where

$$\Theta = \sigma_x + \sigma_y + \sigma_z \tag{2.20}$$

Equation 2.19 can be expressed in tensor form,

$$\nabla^2 \sigma_{\mathrm{ij}} + \frac{1}{1+\nu}\sigma_{\mathrm{kk,ij}} = -\frac{\nu}{1-\nu}\delta_{\mathrm{ij}} div\boldsymbol{X} - \left(X_{i,j} + X_{j,i}\right) \tag{2.21}$$

where

$$\mathrm{div}\mathbf{X} = \nabla \cdot \mathbf{X} = \frac{\partial X_x}{\partial x} + \frac{\partial X_y}{\partial y} + \frac{\partial X_z}{\partial z} \tag{2.22}$$

In an elastic problem, if the equilibrium equations, compatibility equations, and boundary conditions are satisfied, the solution is considered exact.

***Example 2.4*** Suppose a stress field equation is given. How would you determine if it is compatible?

**Solution**

First, substitute the stress field equation into Eq. 2.4 to obtain the body forces *X, Y,* and *Z*. Then, introduce the given stress field equations and the body forces *X, Y,* and *Z* into Eq. 2.19. Finally, check whether the result satisfies Eq. 2.19. If it does not, the given stress field equation is considered incompatible.

## 2.6 Navier's Equilibrium Equation

Navier's equilibrium equation is a form of the equilibrium equation expressed in terms of displacement.

***Example 2.5*** Derive Navier's equilibrium equation from Eq. 2.1.

**Solution**

By introducing Eq. 2.5 into Eq. 2.1,

$$\lambda\delta_{\text{ij}}\varepsilon_{\alpha\alpha.j} + 2G\varepsilon_{\text{ij},j} + X_i = 0\,(a)$$

Introduce $\varepsilon_{\text{ij}} = \frac{1}{2}\left(u_{i,j} + u_{j,i}\right)$ into Equation a,

$$\lambda\delta_{\text{ij}}\left(u_{\alpha,\alpha j}\right) + 2G\frac{1}{2}\left(u_{i,\text{jj}} + u_{j,\text{ij}}\right) + X_i = 0$$

$$G\left(\nabla^2 u_i + u_{j,\text{ji}}\right) + \lambda u_{\alpha,\alpha i} + X_i = 0$$

$$(\because x_i = \delta_{\text{ik}}x_k)$$

$$G\left(\nabla^2 u_i + \frac{\partial\theta}{\partial x_i}\right) + \frac{2Gv}{1-2v}\frac{\partial\theta}{\partial x_i} + X_i = 0$$

$$\left(\theta = u_{\alpha,\alpha} = \varepsilon_x + \varepsilon_y + \varepsilon_z\right) \tag{2.23}$$

$$G\left\{\nabla^2 u_i + \left(1 + \frac{2v}{1-2v}\right)\frac{\partial\theta}{\partial x_i}\right\} + X_i = 0$$

$$\therefore\ G\left(\nabla^2 u_i + \left(\frac{1}{1-2v}\right)\frac{\partial\theta}{\partial x_i}\right) + X_i = 0$$

We call Eq. 2.23 as Navier's equilibrium equation.

# Chapter 3
# Two-Dimensional Elasticity

**Abstract** The difference of plane strain state and plane stress state is stated, and the governing equations of each state are obtained from three-dimensional governing equations.

Two-dimensional elastic problems fall into either a plane strain state or a plane stress state.

## 3.1 Plane Strain State

Plane strain state is defined as follows,

$$
\begin{aligned}
u &= u(x, y) \\
v &= v(x, y) \\
w &= 0
\end{aligned}
\tag{3.1}
$$

In the plane strain state, one component of the displacement vector may not always exist.

Thus, from Eqs. 2.12 and 3.1, we derive Eq. 3.2

$$
\begin{aligned}
\varepsilon_x &= \frac{\partial u}{\partial x} \\
\varepsilon_y &= \frac{\partial v}{\partial y} \\
\varepsilon_{xy} &= \frac{1}{2}\left(\frac{\partial u}{\partial y} + \frac{\partial v}{\partial x}\right) \\
\varepsilon_z &= 0, \varepsilon_{zy} = 0, \varepsilon_{zx} = 0
\end{aligned}
\tag{3.2}
$$

 
K. Y. Lee, *Introduction to Elasticity, Fracture and Fatigue*,
https://doi.org/10.1007/978-981-96-3600-6_3

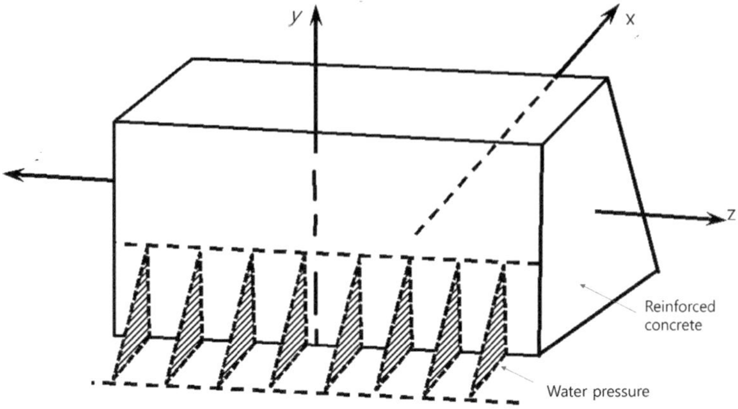


**Fig. 3.1** Dam

Equation 3.2 indicates that a plane strain state occurs when there is no strain on a particular plane. A good example of a plane strain state is a dam, as shown in Fig. 3.1.

**Example 3.1** Demonstrate that the dam shown in Fig. 3.1 satisfies the conditions for a plane strain state.

**Solution**

To confirm that the dam is in a plane strain state, it is necessary to demonstrate that it satisfies Eq. 3.2.

First, let's examine $\varepsilon_z = 0$. Since the dam is constrained on both sides, it cannot be extended in the z-direction. Therefore, $\varepsilon_z$ must be zero. Second, let's check if $\varepsilon_{zy} = 0$ is satisfied. Since the plane defined in the $z$- and $y$-axes cannot distort under the water pressure, $\varepsilon_{zy} = 0$ holds true. Third, let's verify whether $\varepsilon_{zx} = 0$ is satisfied. When examining if the plane defined by the $z$-and $x$-axes can distort, we find distortion cannot occur under water pressure. Thus, $\varepsilon_{zy} = 0$ is satisfied. Therefore, we conclude that the dam is in a plane strain state.

## *3.1.1 Equilibrium Equations*

In the plane strain state, the equilibrium equations derived from Eqs. 2.4 are as follows,

$$
\begin{aligned}
&\frac{\partial \sigma_x}{\partial x} + \frac{\partial \sigma_{\mathrm{xy}}}{\partial y} + X = 0 \\
&\frac{\partial \sigma_{\mathrm{xy}}}{\partial x} + \frac{\partial \sigma_y}{\partial y} + Y = 0 \\
&Z = 0
\end{aligned}
\tag{3.3}
$$

where $X$, $Y$, and $Z$ are body force per unit volume in $x$, $y$ and $z$ axes, respectively.

### 3.1.2 Hooke's Law

Since $\varepsilon_z = 0$ in the plane strain state, from Eqs. 2.8, we have:

$$\varepsilon_z = \frac{1}{E}\left[\sigma_z - \nu\left(\sigma_x + \sigma_y\right)\right] = 0$$

Therefore,

$$\sigma_z = \nu\left(\sigma_x + \sigma_y\right) \tag{3.4}$$

By introducing Eq. 3.4 into the first one of Eq. 2.8,

$$\begin{aligned}\varepsilon_x &= \frac{1}{E}\left[\sigma_x - \nu\left(\sigma_y + \sigma_z\right)\right] = \frac{1}{E}\left[\sigma_x - \nu\left\{\sigma_y + \nu\left(\sigma_x + \sigma_y\right)\right\}\right] \\ &= \frac{1}{E}\left[\left(1 - \nu^2\right)\sigma_x - \nu(1 + \nu)\sigma_y\right]\end{aligned}$$

Therefore,

$$\varepsilon_x = \frac{1 + \nu}{E}\left[(1 - \nu)\sigma_x - \nu\sigma_y\right]$$

Applying a similar process to the second part of Eq. 2.8, we obtain Hooke's law in the plane strain state as follows,

$$\begin{aligned}\varepsilon_x &= \frac{1 + \nu}{E}\left[\sigma_x(1 - \nu) - \nu\sigma_y\right] \\ \varepsilon_y &= \frac{1 + \nu}{E}\left[\sigma_y(1 - \nu) - \nu\sigma_x\right] \\ \varepsilon_z &= 0 \\ \varepsilon_{xy} &= \frac{\sigma_{xy}}{2G} \\ \varepsilon_{xz} &= \varepsilon_{yz} = 0\end{aligned} \tag{3.5}$$

Equation 3.5 is Hooke's law in the plane strain state.

### 3.1.3 Compatibility Equation

Using Eqs. 2.18, the definition of the plane strain state, and the independence of the variable $z$, we find that in the plane strain state, only one compatibility equation is obtained as follows,

$$2\frac{\partial^2 \varepsilon_{xy}}{\partial x \partial y} = \frac{\partial^2 \varepsilon_x}{\partial y^2} + \frac{\partial^2 \varepsilon_y}{\partial x^2} \tag{3.6}$$

Equation 3.6 is a compatibility equation in the plain strain state.

### 3.1.4 Beltrami-Michell Compatibility Equation

By applying Eqs. 3.3 and 3.5 to Eq. 3.6, we can derive Eq. 3.7.

$$\nabla^2(\sigma_x + \sigma_y) = -\frac{1}{1-\nu}\left(\frac{\partial X}{\partial x} + \frac{\partial Y}{\partial y}\right) \tag{3.7}$$

This is called Beltrami-Michell compatibility equation in the plane strain state.

### 3.1.5 Traction Vector

By applying the plane strain state definition and setting $n_z = 0$ into Eq. 2.14, we obtain Eq. 3.8.

$$\begin{aligned} T_x &= \sigma_x n_x + \sigma_{xy} n_y \\ T_y &= \sigma_{xy} n_x + \sigma_y n_y \\ T_z &= 0 \end{aligned} \tag{3.8}$$

This is the traction vector in the plane strain state.

### 3.1.6 Navier's Equilibrium Equations

Following a detailed derivation using Eqs. 3.2, 3.3, and 3.5, we obtain Eq. 3.9.

$$
\begin{aligned}
G\nabla^2 u + \frac{G}{1-2v}\frac{\partial}{\partial x}\left(\frac{\partial u}{\partial x} + \frac{\partial v}{\partial y}\right) + X = 0 \\
G\nabla^2 v + \frac{G}{1-2v}\frac{\partial}{\partial y}\left(\frac{\partial u}{\partial x} + \frac{\partial v}{\partial y}\right) + Y = 0
\end{aligned}
\tag{3.9}
$$

where $\nabla^2 = \frac{\partial^2}{\partial x^2} + \frac{\partial^2}{\partial y^2}$.

Equation 3.9 is called Navier's equilibrium equations in the plane strain state.

## 3.2 Plane Stress State

When there is no stress on one of the planes, as shown in Eq. 3.10, the material is said to be in a plane stress state.

$$
\begin{aligned}
&\sigma_z = 0, \sigma_{zx} = 0, \sigma_{zy} = 0 \\
&\sigma_x = \sigma_x(x, y), \sigma_y = \sigma_y(x, y), \sigma_{xy} = \sigma_{xy}(x, y)
\end{aligned}
\tag{3.10}
$$

As a good example for the plane stress state, the thin circular plate which is loaded by uniform pressure on the external boundary is shown in Fig. 3.2.

**Example 3.2** Test whether the plate shown in Fig. 3.2 satisfies the condition of plane stress state.

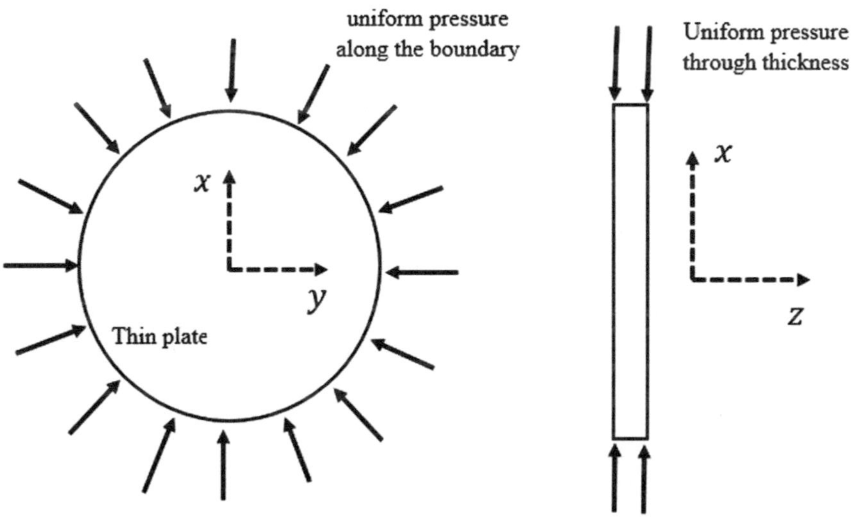


**Fig. 3.2** Circular plate under uniform pressure on the external boundary

**Solution**

For the plate to be in a plane stress state, it must satisfy Eq. 3.10. First, let's examine $\sigma_z = 0$. The $z$-axis is toward thickness direction. Both sides of the plate are stress-free. Therefore, they satisfy $\sigma_z = 0$ there. However, in the middle surface of thickness, $\sigma_z$ is not zero although it has a little stress value because of compressive pressure on the external boundary. But $\sigma_{zx} = 0$ is satisfied because the plane surrounded in the z-and x-axes cannot be distorted under uniform compressive pressure on the external boundary. $\sigma_{zy} = 0$ is also satisfied because the plane surrounded in the z- and y-axes cannot be distorted. Therefore, Eq. 3.10 is not perfectly satisfied because $\sigma_z$ is not zero on all $z$-plane but approximately zero. As demonstrated in this example, the plane stress state cannot be perfectly achieved but can be approximated in two-dimensional elastic problems.

### 3.2.1 *Equilibrium Equation*

The equilibrium equations for the plane stress state, derived from Eqs. 2.4 and 3.10, are as follows,

$$\begin{aligned}\frac{\partial\sigma_x}{\partial x}+\frac{\partial\sigma_{\mathrm{xy}}}{\partial y}+X&=0\\ \frac{\partial\sigma_{\mathrm{xy}}}{\partial x}+\frac{\partial\sigma_y}{\partial y}+Y&=0\\ Z&=0\end{aligned}\tag{3.11}$$

It is noted that equilibrium equations are the same in both plane strain and plane stress states.

### 3.2.2 *Hooke's Law*

By substituting Eq. 3.10 into Eq. 2.8, we obtain Eq. 3.12, which represents Hooke's law in the plane stress state.

$$\begin{aligned}\varepsilon_x&=\frac{1}{E}\left(\sigma_x-\nu\sigma_y\right)\\ \varepsilon_y&=\frac{1}{E}\left(\sigma_y-\nu\sigma_x\right)\\ \varepsilon_z&=-\frac{\nu}{E}\left(\sigma_x+\sigma_y\right)\\ \varepsilon_{\mathrm{xy}}&=\frac{\sigma_{\mathrm{xy}}}{2G}\end{aligned}$$

$$\varepsilon_{xz} = 0$$
$$\varepsilon_{yz} = 0 \tag{3.12}$$

Note that Hooke's law differs between the plane strain state and the plane stress state.

Equation 3.13 is another form of Hooke's law derived from Eq. 3.12.

$$\begin{aligned}
\sigma_x &= \frac{E}{1-\nu^2}\left(\varepsilon_x + \nu\varepsilon_y\right)\\
\sigma_y &= \frac{E}{1-\nu^2}\left(\varepsilon_y + \nu\varepsilon_x\right)\\
\sigma_{xy} &= 2G\varepsilon_{xy}\\
\sigma_z &= \sigma_{zx} = \sigma_{yz} = 0
\end{aligned} \tag{3.13}$$

### 3.2.3 Compatibility Equation

Using Eq. 3.10 and the fact that strains are not functions of $z$ in a two-dimensional problem, we derive Eq. 3.14, corresponding to Eq. 2.18, as follows:

$$\begin{aligned}
&0 = 0\\
&0 = 0\\
&\frac{\partial^2 \varepsilon_z}{\partial x \partial y} = 0\\
&2\frac{\partial^2 \varepsilon_{xy}}{\partial x \partial y} = \frac{\partial^2 \varepsilon_x}{\partial y^2} + \frac{\partial^2 \varepsilon_y}{\partial x^2}\\
&0 = 0 + \frac{\partial^2 \varepsilon_z}{\partial y^2}\\
&0 = \frac{\partial^2 \varepsilon_z}{\partial x^2}
\end{aligned} \tag{3.14}$$

But, we pick up only Eq. 3.15 from Eq. 3.14 as a compatibility equation

$$2\frac{\partial^2 \varepsilon_{xy}}{\partial x \partial y} = \frac{\partial^2 \varepsilon_x}{\partial y^2} + \frac{\partial^2 \varepsilon_y}{\partial x^2} \tag{3.15}$$

Therefore, the solution to the elastic problem in the plane stress state is not an exact solution, but it provides a good approximation. However, it is noted that the compatibility equation written in terms of strain in both plane strain and plane stress states is same.

### 3.2.4 Beltrami-Michell Compatibility Equation

Equation 3.16 is derived by substituting Eqs. 3.11 and 3.12 into Eq. 3.15.

$$\nabla^2\left(\sigma_x+\sigma_y\right)=-(1+\nu)\left(\frac{\partial X}{\partial x}+\frac{\partial Y}{\partial y}\right) \tag{3.16}$$

This is called Beltrami-Michell compatibility equation in the plane stress state.

Note that the Beltrami-Michell compatibility equations differ between the plane strain and plane stress states.

**Example 3.3** Although the compatibility equation written in terms of strain in both plane strain and plane stress states are the same, why are Beltrami-Michell compatibility equations in the plane strain state and plane stress state different?

**Solution**

In both states, the same equilibrium equations and compatibility equation are used to derive the Beltrami-Michell compatibility equations. However, because Hooke's law differs between the two states, the resulting Beltrami-Michell compatibility equations also differ.

### 3.2.5 Traction Vector

The traction vector in the plane stress state is obtained from Eqs. 2.14 and 3.10 as follows,

$$\begin{aligned} T_x &= \sigma_x n_x + \sigma_{\mathrm{xy}} n_y \\ T_y &= \sigma_{\mathrm{xy}} n_x + \sigma_y n_y \\ T_z &= 0 \end{aligned} \tag{3.17}$$

### 3.2.6 Navier's Equilibrium Equations

Equation 3.11 can be expressed in terms of displacement resulting in Eq. 3.18, known as Navier's equilibrium equations.

$$\begin{aligned} G\nabla^2 u + \frac{E}{2(1-\nu)}\frac{\partial}{\partial x}\left(\frac{\partial u}{\partial x}+\frac{\partial v}{\partial y}\right) + X = 0 \\ G\nabla^2 v + \frac{E}{2(1-\nu)}\frac{\partial}{\partial y}\left(\frac{\partial u}{\partial x}+\frac{\partial v}{\partial y}\right) + Y = 0 \end{aligned} \tag{3.18}$$

Equation 3.18 can be derived by using the equilibrium equations, Hooke's law corresponding to the plane stress state and the relationship between strain and displacement.

### 3.2.7 *Replacement from One State to Another State*

By substituting $E$ and $\nu$ in the plane strain state equations with $E_1$ and $\nu_1$ from Eq. 3.19, the corresponding plane stress state equations can be obtained.

$$E_1 = \frac{E(1+2\nu)}{(1+\nu)^2}, \quad \nu_1 = \frac{\nu}{1+\nu} \tag{3.19}$$

**Example 3.4** Using Eq. 3.19, derive the Beltrami-Michell compatibility equations for the plane stress state from those in the plane strain state.

**Solution**

Beltrami-Michell compatibility equations in the plane strain state is as follows as shown in Eq. 3.7,

$$\nabla^2(\sigma_x + \sigma_y) = -\frac{1}{1-\nu}\left(\frac{\partial X}{\partial x} + \frac{\partial Y}{\partial y}\right)$$

By using Eq. 3.19,

$$\begin{aligned} \nabla^2(\sigma_x + \sigma_y) &= -\frac{1}{1-\frac{\nu}{1+\nu}}\left(\frac{\partial X}{\partial x} + \frac{\partial Y}{\partial y}\right) \\ &= -\frac{1+\nu}{1+\nu-\nu}\left(\frac{\partial X}{\partial x} + \frac{\partial Y}{\partial y}\right) \end{aligned}$$

Therefore, we can get Beltrami-Michell compatibility equations in the plane stress state as follows as shown in Eq. 3.16,

$$\nabla^2(\sigma_x + \sigma_y) = -(1+\nu)\left(\frac{\partial X}{\partial x} + \frac{\partial Y}{\partial y}\right)$$

**Example 3.5** Using Eq. 3.19, derive the Navier equilibrium equation for the plane stress state from the equation in the plane strain state.

**Solution**

The first of Navier's equilibrium equations in the plane strain state is given in Eq. 3.9,

$$G\nabla^2 u + \frac{G}{1-2\nu}\frac{\partial}{\partial x}\left(\frac{\partial u}{\partial x} + \frac{\partial v}{\partial y}\right) + X = 0$$

By applying Eq. 3.19, we obtain the following,

$$\frac{E\frac{(1+2\nu)}{(1+\nu)^2}}{2\left(1+\frac{\nu}{1+\nu}\right)}\left[\nabla^2 u+\frac{1}{1-2\frac{\nu}{1+\nu}}\frac{\partial}{\partial x}\left(\frac{\partial u}{\partial x}+\frac{\partial v}{\partial y}\right)\right]+X=0$$

Finally, we derive the Navier's equilibrium equation in the plane stress state, as shown in the first part of Eq. 3.18,

$$G\nabla^2 u+\frac{E}{2(1-\nu)}\frac{\partial}{\partial x}\left(\frac{\partial u}{\partial x}+\frac{\partial v}{\partial y}\right)+X=0$$

By substituting $E_2$ and $\nu_2$ (as defined below) for $E$ and $\nu$ in the equations for the plane stress state, we obtain the corresponding plane strain state equations.

$$E_2=\frac{E}{1-\nu^2},\quad \nu_2=\frac{\nu}{1-\nu} \tag{3.20}$$

Therefore, by using Eq. 3.20, we can also solve a two-dimensional elastic problem under the plane strain state by treating it as the problem under the plane stress state.

**Example 3.6** Using Eq. 3.20, derive Hooke's law for the plane strain state from Hooke's law given for the plane stress state.

**Solution**

Hooke's law in the plane stress state is presented in Eq. 3.12. One component is given as follows,

$$\varepsilon_x=\frac{1}{E}\left(\sigma_x-\nu\sigma_y\right)$$

Applying Eq. 3.20, we have: $\varepsilon_x=\frac{1}{E}\left[\sigma_x-\nu\sigma_y\right]=\frac{1-\nu^2}{E}\left[\sigma_x-\frac{\nu}{1-\nu}\sigma_y\right]$

$$\therefore\varepsilon_x=\frac{1+\nu}{E}\left[(1-\nu)\sigma_x-\nu\sigma_y\right]$$

This result matches the first equation of Hooke's law in the plane strain state, as shown in Eq. 3.5.

# Chapter 4
# Airy Stress Function

**Abstract** The Airy stress functions in Cartesian and polar coordinates are introduced. The transformation method between the governing equations in Cartesian coordinate and polar coordinate is also explained.

## 4.1 Airy Stress Function in Cartesian Coordinate

The Airy stress function $\phi(x, y)$ is defined as follows,

$$\sigma_x + \Omega = \frac{\partial^2 \phi}{\partial y^2}, \quad \sigma_y + \Omega = \frac{\partial^2 \phi}{\partial x^2}$$
$$\sigma_{xy} = -\frac{\partial^2 \phi}{\partial x \partial y} \tag{4.1}$$

where

$$X = \frac{\partial \Omega}{\partial x}, \quad Y = \frac{\partial \Omega}{\partial y} \tag{4.2}$$

Here, $\Omega$ is a potential function representing an virtual function. When Eq. 4.1 is substituted into the equilibrium equations (Eqs. 3.3 and 3.11), it can be shown that Eq. 4.1 automatically satisfies these equilibrium equations.

***Example 4.1*** Show that Eq. 4.1 automatically satisfies Eqs. 3.3 and 3.11.

**Solution**

Note that Eqs. 3.3 and 3.11 are identical. The first equilibrium equation can be written as,

$$\frac{\partial \sigma_x}{\partial x} + \frac{\partial \sigma_{xy}}{\partial y} + X = 0 \tag{4.3}$$

 
K. Y. Lee, *Introduction to Elasticity, Fracture and Fatigue*,
https://doi.org/10.1007/978-981-96-3600-6_4

By introducing Eq. 4.1 into Eq. 4.3, we get,

$$\frac{\partial}{\partial x}\left(\frac{\partial^2\phi}{\partial y^2}-\Omega\right)+\frac{\partial}{\partial y}\left(-\frac{\partial^2\phi}{\partial x\partial y}\right)+\frac{\partial\Omega}{\partial x}=0$$

Thus, we confirm that Eq. 4.1 satisfies the first equilibrium equation. Similarly, we can verify that it also satisfies the second equilibrium equation. Therefore, when the Airy stress function is defined as in Eq. 4.1, it satisfies the equilibrium equations.

Compatibility equations in terms of the Airy stress function in both the plane strain and plane stress states are given by

$$\nabla^4\phi=\left(\frac{1-2\nu}{1-\nu}\right)\nabla^2\Omega \quad \text{(plane strain state)} \tag{4.4}$$

$$\nabla^4\phi=(1-\nu)\nabla^2\Omega \quad \text{(plane stress state)} \tag{4.5}$$

where

$$\nabla^4\equiv\frac{\partial^4}{\partial x^4}+2\frac{\partial^4}{\partial x^2\partial y^2}+\frac{\partial^4}{\partial y^4},\quad \nabla^2=\frac{\partial^2}{\partial x^2}+\frac{\partial^2}{\partial y^2} \tag{4.6}$$

***Example 4.2*** Using the Beltrami-Michell compatibility equation for the plane stress state (Eq. 3.16) and Eq. 4.1, show that the compatibility equation in terms of the Airy stress function can be written as in Eq. 4.5.

**Solution**

The Beltrami-Michell compatibility equation in the plane stress state (Eq. 3.16) is given by,

$$\nabla^2(\sigma_x+\sigma_y)=-(1+\nu)\left(\frac{\partial X}{\partial x}+\frac{\partial Y}{\partial y}\right) \tag{4.7}$$

Substituting Eqs. 4.1 and 4.2 into Eq. 4.3 results in,

$$\nabla^2\left(\frac{\partial^2\phi}{\partial y^2}-\Omega+\frac{\partial^2\phi}{\partial x^2}-\Omega\right)=-(1+\nu)\left(\frac{\partial}{\partial x}\left(\frac{\partial\Omega}{\partial x}\right)+\frac{\partial}{\partial y}\left(\frac{\partial\Omega}{\partial y}\right)\right)$$

Thus, we arrive at,

$$\nabla^4\phi=(1-\nu)\nabla^2\Omega$$

If $\nabla^2\Omega=0$, assuming no body forces, Eqs. 4.4 and 4.5 simplify to,

$$\nabla^4\phi=0 \tag{4.8}$$

***Example 4.3*** Using Timoshenko's beam theory, determine the stresses in a cantilever beam under a shear force $P$ per unit thickness at $x = 0$, based on the Airy stress function. The beam has unit thickness in the $z$-direction, length $L$ in $x$-direction and height 2a in $y$-direction. Assume the body forces are neglected. (Reference: A. C. Ugural and S. K. Fenster, Advanced Strength and Applied Elasticity, Elsevier North Holland Publishing Co., Inc., pp. 71–73, 1978).

**Solution**

It is noted that the following boundary conditions are obtained;

1. at the upper and low surface $y = \pm a, \quad \sigma_y = 0, \sigma_{xy} = 0,$
2. at $x = 0, \sigma_x = 0, \int_{-a}^{a} \sigma_{xy} \mathrm{d}y + P = 0$
3. at any $x$, $\int_{-a}^{a} \sigma_x \mathrm{d}y = 0$

Now, try to get the Airy stress function which satisfies $\nabla^4 \phi = 0$ from boundary conditions.

Timoshenko's beam theory is written as follows,

$$\sigma_{xy} = f_1(y) \tag{4.9}$$

It is important to know what Eq. 4.9 means.

That means that since $\sigma_{xy}$ is not a function of $x$, $\sigma_{xy}$ at the locations $a$, $b$, $c$, and $d$ shown in Fig. 4.1 are all same.

From the definition of Airy stress function, Eq. 4.9 is written as follows,

$$\sigma_{xy} = -\frac{\partial^2 \phi}{\partial x \partial y} = f_1(y) \tag{4.10}$$

By integrating Eq. 4.10 with respect to $x$ and $y$, we can get Eqs. 4.11 and 4.12,

$$-\frac{\partial \phi}{\partial x} = \int f_1(y) \mathrm{d}y + f_2(x) \tag{4.11}$$

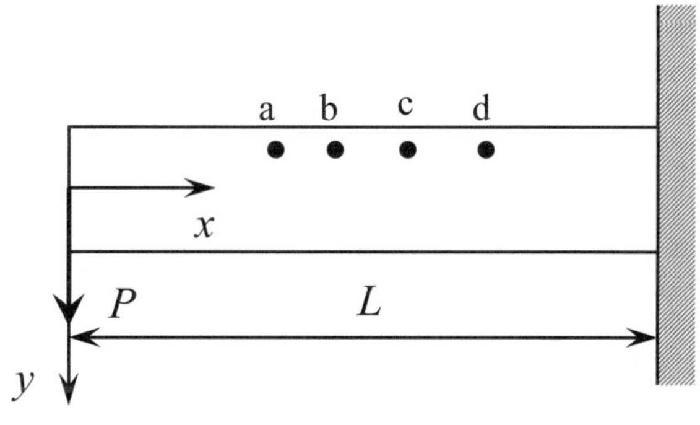


**Fig. 4.1** Timoshenko's beam theory

$$-\phi = x\int f_1(y)\mathrm{d}y + \int f_2(x)\mathrm{d}x + f_3(y) \tag{4.12}$$

Since Eq. 4.10 shows partial derivative, it is noted that integral constants, $f_2(x)$ and $f_3(y)$ should be as follows of a function. Those should be determined by boundary conditions.

From Eq. 4.11

$$\sigma_y = \frac{\partial^2\phi}{\partial x^2} = -\frac{\mathrm{d}f_2(x)}{\mathrm{d}x} \tag{4.13}$$

It is noted that since $f_2(x)$ is only in the function of $x$, Eq. 4.13 is expressed in the total derivative form.

From Eq. 4.13 and $\sigma_y = 0$ at the upper and low surface surface $y = \pm a$,

$$f_2(x) = A \tag{4.14}$$

By introducing Eq. 4.14 into Eq. 4.12,

$$-\phi = x\int f_1(y)\mathrm{d}y + Ax + f_3(y) \tag{4.15}$$

Since $\phi$ in Eq. 4.1 has the second order of derivative, constant terms and the linear terms with respect to $x$ and $y$ in $\phi$ do not give any effect on stresses. Therefore, By disregarding $Ax$ in Eq. 4.15

$$-\phi = x\int f_1(y)\mathrm{d}y + f_3(y) \tag{4.16}$$

From the definition of Airy stress function and Eq. 4.16,

$$\sigma_x = \frac{\partial^2\phi}{\partial y^2} = -x\frac{\mathrm{d}f_1(y)}{\mathrm{d}y} - \frac{d^2f_3(y)}{\mathrm{d}y^2} \tag{4.17}$$

From the boundary condition $\sigma_x = 0$ at $x = 0$,

$$\frac{d^2f_3(y)}{\mathrm{d}y^2} = 0 \tag{4.18}$$

Therefore,

$$f_3(y) = cy + D \tag{4.19}$$

where $C$ and $D$ are real constant.

By introducing Eq. 4.19 into Eq. 4.16,

$$-\phi = x\int f_1(y)\mathrm{d}y + Cy + D$$

By disregarding $Cy + D$ terms,

$$\phi = -x\int f_1(y)\mathrm{d}y \tag{4.20}$$

Since $f_1(y)$ is not an integral constant as shown in Eq. 4.9, it should be determined by a governing equation. The governing equation is only Eq. 4.8.

Equation 4.8 is as follows,

$$\nabla^4\phi = \left(\frac{\partial^2}{\partial x^2} + \frac{\partial^2}{\partial y^2}\right)\left(\frac{\partial^2}{\partial x^2} + \frac{\partial^2}{\partial y^2}\right)\phi = 0 \tag{4.21}$$

By introducing Eq. 4.20 into Eq. 4.21,

$$-x\frac{d^3 f_1(y)}{\mathrm{d}y^3} = 0 \tag{4.22}$$

Therefore,

$$f_1(y) = \frac{E}{2}y^2 + Fy + G \tag{4.23}$$

where $E$, $F$ and $G$ are real constant as integral constants.

By introducing Eq. 4.23 into Eq. 4.20

$$\phi = x\int\left(\frac{\alpha}{2}y^2 + \beta y + \gamma\right)\mathrm{d}y \tag{4.24}$$

where $\alpha = -E,\quad \beta = -F,\quad \gamma = -G.$

By the definition of Airy stress function and Eq. 4.24,

$$\sigma_x = \frac{\partial^2\phi}{\partial y^2} = \frac{\partial^2}{\partial y^2}\left[x\int\left(\frac{\alpha}{2}y^2 + \beta y + \gamma\right)\mathrm{d}y\right] = \alpha xy + x\beta \tag{4.25}$$

From the boundary condition 3 and Eq. 4.25,

$$\int_{-a}^{a}(\alpha xy + x\beta)\mathrm{d}y = 0 \tag{4.26}$$

Therefore,

$$\alpha x \frac{y^2}{2}\bigg|_{-a}^{a} + x\beta y|_{-a}^{a} = 0 \tag{4.27}$$

Therefore,

$$\beta = 0 \tag{4.28}$$

By introducing Eq. 4.28 into Eq. 4.24,

$$\phi = x \int \left(\frac{\alpha}{2} y^2 + \gamma\right) \mathrm{d}y \tag{4.29}$$

Therefore,

$$\sigma_{xy} = -\frac{\partial^2 \phi}{\partial x \partial y} = -\left(\frac{\alpha}{2} y^2 + \gamma\right) \tag{4.30}$$

By the boundary condition $\sigma_{xy} = 0$ at $y = \pm a$ and Eq. 4.30,

$$\gamma = -\frac{\alpha}{2} a^2 \tag{4.31}$$

By introducing Equation 4.31 into Equation 4.30,

$$\sigma_{xy} = -\frac{\alpha}{2}\left(y^2 - a^2\right) \tag{4.32}$$

By the boundary condition $\int_{-a}^{a} \sigma_{xy} \mathrm{d}y = -P$ at $x = 0$,

$$-P = \int_{-a}^{a} \sigma_{xy} \mathrm{d}y = \int_{-a}^{a} \frac{\alpha}{2}\left(a^2 - y^2\right) \mathrm{d}y$$

$$-P = \frac{\alpha}{2} a^2 y \bigg|_{-a}^{a} - \frac{\alpha}{2} \frac{y^3}{3}\bigg|_{-a}^{a}$$

Therefore,

$$\alpha = -\frac{3P}{2a^3} \tag{4.33}$$

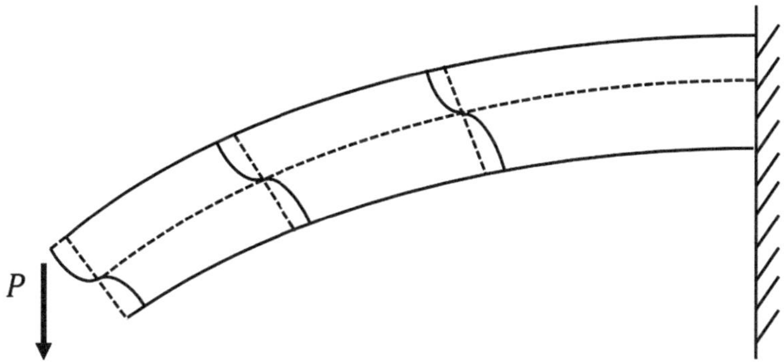


**Fig. 4.2** Warping of the cross section of a beam due to shear stress

Finally, by introducing Eqs. 4.31 and 4.33 into Eq. 4.29,

$$\phi = -\frac{xPy^3}{4a^3} + \frac{3Pxy}{4a} \tag{4.34}$$

By introducing Eq. 4.34 into the definition of Airy stress function, finally we get,

$$\begin{aligned}
\sigma_x &= -\frac{3Pxy}{2a^3} \\
\sigma_y &= 0 \\
\sigma_{xy} &= \frac{3P}{4a^3}\left(y^2 - a^2\right)
\end{aligned}$$

It is important to realize that the cross section of the beam shows warping as shown in Fig. 4.2 due to shear stress.

## 4.2 Polar Coordinate

The polar coordinate is a two-dimensional coordinate system, as shown in Fig. 4.3 From Fig. 4.3,

$$\begin{aligned}
x &= r\cos\theta \\
y &= r\sin\theta \\
\tan\theta &= \frac{y}{x}
\end{aligned} \tag{4.35}$$

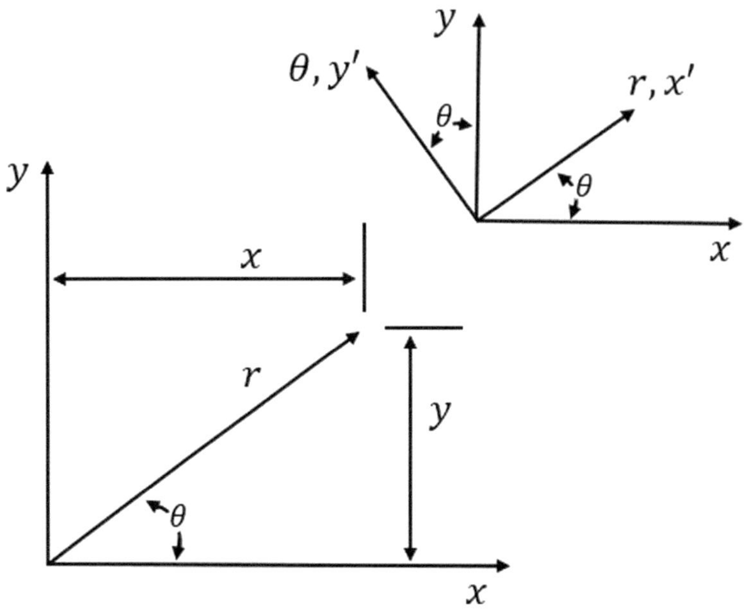


**Fig. 4.3** Cartesian and polar coordinates

Therefore,

$$r^2 = x^2 + y^2$$
$$\theta = \tan^{-1}\frac{y}{x} \tag{4.36}$$

When we take the partial derivatives with respect to $x$ and $y$ for the first of Eq. 4.36, we have,

$$2r\frac{\partial r}{\partial x} = 2x$$
$$2r\frac{\partial r}{\partial y} = 2y$$

Thus, we find,

$$\frac{\partial r}{\partial x} = \frac{x}{r} = \frac{r\cos\theta}{r} = \cos\theta$$
$$\frac{\partial r}{\partial y} = \frac{y}{r} = \frac{r\sin\theta}{r} = \sin\theta$$

When we take the partial derivative to the second one of Eq. 4.36 with respect of $x$ and $y$ by using the formula as follows,

$$\frac{\mathrm{d}}{\mathrm{d}x}\tan^{-1}u = \frac{1}{1+u^2}\frac{\mathrm{d}u}{\mathrm{d}x}$$

we can get

$$\begin{aligned}\frac{\partial\theta}{\partial x} &= \frac{-\sin\theta}{r}\\ \frac{\partial\theta}{\partial y} &= \frac{\cos\theta}{r}\end{aligned}$$

Therefore,

$$\begin{aligned}&\frac{\partial r}{\partial x} = \cos\theta,\ \frac{\partial r}{\partial y} = \sin\theta\\ &\frac{\partial\theta}{\partial x} = -\frac{\sin\theta}{r},\ \frac{\partial\theta}{\partial y} = \frac{\cos\theta}{r}\end{aligned} \tag{4.37}$$

By applying Eq. 4.37 into the chain rule for derivative as follows,

$$\frac{\partial f(x,y)}{\partial x} = \frac{\partial f(x,y)}{\partial r}\frac{\partial r}{\partial x} + \frac{\partial f(x,y)}{\partial\theta}\frac{\partial\theta}{\partial x}$$

we obtain,

$$\begin{aligned}\frac{\partial f(x,y)}{\partial x} &= \cos\theta\frac{\partial f(r,\theta)}{\partial r} - \frac{\sin\theta}{r}\frac{\partial f(r,\theta)}{\partial\theta}\\ \frac{\partial f(x,y)}{\partial y} &= \sin\theta\frac{\partial f(r,\theta)}{\partial r} + \frac{\cos\theta}{r}\frac{\partial f(r,\theta)}{\partial\theta}\end{aligned} \tag{4.38}$$

These formulas (4.38) are used to convert partial derivatives with respect to $x$ and $y$ into partial derivatives with respect to $r$ and $\theta$.

***Example 4.4*** Derive the relationship between strain and displacement in polar coordinates using Eq. 4.38.

**Solution**

To derive the relationship between strain and displacement in polar coordinates, we follow several steps,

Step 1. Based on stress transformation, the strain transformation can be expressed as,

$$\varepsilon'_{lm} = a_{li}a_{mj}\varepsilon_{ij} \tag{4.39}$$

where $l, m = x', y', z'$ and $i, j = x, y, z$.

Step 2. To find strain transformation in polar coordinates, substitute $x'$, $y'$ into $r$, $\theta$ in Eq. 4.39, yielding,

$$\varepsilon_r = a_{rx}^2\varepsilon_x + a_{ry}^2\varepsilon_y + 2a_{rx}a_{ry}\varepsilon_{xy}$$

By expressing the direction cosines as the terms of cosine,

$$\begin{aligned}\varepsilon_r &= a_{rx}^2\varepsilon_x + a_{ry}^2\varepsilon_y + 2a_{rx}a_{ry}\varepsilon_{xy}\\ &= \cos^2(r,x)\varepsilon_x + \cos^2(r,y)\varepsilon_y + 2\cos(r,x)\cos(r,y)\varepsilon_{xy}\\ &= \varepsilon_x\cos^2\theta + \cos^2(90-\theta)\varepsilon_y + 2\cos\theta\cos(90-\theta)\varepsilon_{xy}\\ &= \varepsilon_x\cos^2\theta + \varepsilon_y\sin^2\theta + \varepsilon_{xy}\sin 2\theta\end{aligned}$$

By applying the similar processes to $\varepsilon_\theta$ and $\varepsilon_{r\theta}$,

$$\begin{aligned}\varepsilon_r &= \varepsilon_x\cos^2\theta + \varepsilon_y\sin^2\theta + \varepsilon_{xy}\sin 2\theta\\ \varepsilon_\theta &= \varepsilon_x\sin^2\theta + \varepsilon_y\cos^2\theta - \varepsilon_{xy}\sin 2\theta\\ \varepsilon_{r\theta} &= (\varepsilon_y - \varepsilon_x)\cos\theta\sin\theta + \varepsilon_{xy}\left(\cos^2\theta - \sin^2\theta\right)\end{aligned} \tag{4.40}$$

Step 3. Since displacement transformation is shown as

$$u_i = \beta_{ij}u_j' \tag{4.41}$$

where $\beta_{ij}$ is direction cosine, we can write,

$$u_x = \beta_{xx'}u_x' + \beta_{xy'}u_y' \tag{4.42}$$

By changing the index $x'$ and $y'$ into $r$ and $\theta$ in Eq. 4.42 and expressing the direction cosines as the terms of cosine,

$$\begin{aligned}u_x &= \cos(x,r)u_r + \cos(x,\theta)u_\theta\\ &= u_r\cos\theta + u_\theta\cos(90+\theta) = u_r\cos\theta - u_\theta\sin\theta\end{aligned}$$

Finally, we can express the displacement transformations as follows,

$$\begin{aligned}u_x &= u_r\cos\theta - u_\theta\sin\theta\\ u_y &= u_r\sin\theta + u_\theta\cos\theta\end{aligned} \tag{4.43}$$

Substituting Eq. 4.43 into the relationship between strain and displacement in Cartesian coordinates, and applying Eq. 4.38, we obtain,

$$\begin{aligned}\varepsilon_x &= \frac{\partial u_x}{\partial x} = \left(\cos\theta\frac{\partial}{\partial r} - \frac{\sin\theta}{r}\frac{\partial}{\partial\theta}\right)(u_r\cos\theta - u_\theta\sin\theta)\\ &= \cos^2\theta\frac{\partial u_r}{\partial r} + \sin^2\theta\left(\frac{u_r}{r} + \frac{1}{r}\frac{\partial u_\theta}{\partial\theta}\right) - \cos\theta\sin\theta\left(\frac{\partial u_\theta}{\partial r} + \frac{\partial u_r}{r\partial\theta} - \frac{u_\theta}{r}\right)\end{aligned}$$

$$\begin{aligned}
\varepsilon_y &= \frac{\partial u_y}{\partial y} \\
&= \sin^2\theta\frac{\partial u_r}{\partial r} + \cos^2\theta\left(\frac{u_r}{r} + \frac{\partial u_\theta}{r\partial\theta}\right) + \cos\theta\sin\theta\left(\frac{\partial u_\theta}{\partial r} + \frac{\partial u_r}{r\partial\theta} - \frac{u_\theta}{r}\right) \\
\varepsilon_{xy} &= \frac{1}{2}\left(\frac{\partial u_x}{\partial y} + \frac{\partial u_y}{\partial x}\right) \\
&= \frac{\sin 2\theta}{2}\left(\frac{\partial u_r}{\partial r} - \frac{\partial u_\theta}{r\partial\theta} - \frac{u_r}{r}\right) + \frac{\cos 2\theta}{2}\left(\frac{\partial u_\theta}{\partial r} + \frac{\partial u_r}{r\partial\theta} - \frac{u_\theta}{r}\right)
\end{aligned} \tag{4.44}$$

By introducing Eq. 4.44 into Eq. 4.40,

$$\begin{aligned}
\varepsilon_r &= \frac{\partial u_r}{\partial r} \\
\varepsilon_\theta &= \frac{u_r}{r} + \frac{1}{r}\frac{\partial u_\theta}{\partial\theta} \\
\varepsilon_{r\theta} &= \frac{1}{2}\left(\frac{1}{r}\frac{\partial u_r}{\partial\theta} + \frac{\partial u_\theta}{\partial r} - \frac{u_\theta}{r}\right)
\end{aligned} \tag{4.45}$$

***Example 4.5*** Derive the compatibility equation in polar coordinates.

**Solution**

The three-dimensional compatibility equation, Eq. 2.17, was obtained by eliminating displacements in Eq. 2.15. By applying the same method to Eq. 4.45, we obtain the compatibility equation in polar coordinates as follows,

$$\frac{\partial^2\varepsilon_\theta}{\partial r^2} + \frac{1}{r^2}\frac{\partial^2\varepsilon_r}{\partial\theta^2} + \frac{2}{r}\frac{\partial\varepsilon_\theta}{\partial r} - \frac{1}{r}\frac{\partial\varepsilon_r}{\partial r} = \frac{2}{r}\frac{\partial^2\varepsilon_{r\theta}}{\partial r\partial\theta} + \frac{2}{r^2}\frac{\partial\varepsilon_{r\theta}}{\partial\theta} \tag{4.46}$$

***Example 4.6*** Derive Hooke's law in polar coordinates.

**Solution**

By replacing the indices $x$ and $y$ in Hooke's law (Eqs. 3.5 and 3.12 in Cartesian coordinates) with $r$ and $\theta$, we can obtain Hooke's law in plane strain state as follows,

$$\begin{aligned}
\varepsilon_r &= \frac{1+\nu}{E}[\sigma_r(1-\nu) - \nu\sigma_\theta] \\
\varepsilon_\theta &= \frac{1+\nu}{E}[\sigma_\theta(1-\nu) - \nu\sigma_r] \\
\varepsilon_{r\theta} &= \frac{\sigma_{r\theta}}{2G}
\end{aligned} \tag{4.47}$$

and in plane stress state as follows,

$$\varepsilon_r = \frac{1}{E}[\sigma_r - \nu\sigma_\theta]$$

$$\varepsilon_\theta = \frac{1}{E}[\sigma_\theta - \nu\sigma_r]$$
$$\varepsilon_{r\theta} = \frac{\sigma_{r\theta}}{2G} \tag{4.48}$$

***Example 4.7*** Convert $\nabla^4\phi(x, y) = 0$ to its corresponding form in polar coordinates using Eq. 4.38.

**Solution**

$$\nabla^2 = \frac{\partial^2}{\partial x^2} + \frac{\partial^2}{\partial y^2} \tag{4.49}$$

By introducing Eq. 4.38 into Eq. 4.49,

$$\nabla^4\phi = \nabla^2\nabla^2\phi = \left(\frac{\partial^2}{\partial r^2} + \frac{1}{r}\frac{\partial}{\partial r} + \frac{1}{r^2}\frac{\partial^2}{\partial\theta^2}\right)\left(\frac{\partial^2}{\partial r^2} + \frac{1}{r}\frac{\partial}{\partial r} + \frac{1}{r^2}\frac{\partial^2}{\partial\theta^2}\right)\phi(r,\theta) = 0 \tag{4.50}$$

## 4.3 Airy Stress Function in Polar Coordinate

Given the definition of the Airy stress function in Cartesian coordinates, we can derive the equations relating stress and the Airy stress function in polar coordinates using Eq. 4.38.

***Example 4.8*** Using the definition of the Airy stress function in Cartesian coordinates, derive the equations relating stress and the Airy stress function in polar coordinates with the help of Eq. 4.38.

**Solution**

Using stress transformation, or by substituting $\varepsilon$ with $\sigma$ in Eq. 4.40,

$$\sigma_r = \sigma_x\cos^2\theta + \sigma_y\sin^2\theta + \sigma_{xy}\sin 2\theta$$
$$\sigma_\theta = \sigma_x\sin^2\theta + \sigma_y\cos^2\theta - \sigma_{xy}\sin 2\theta$$
$$\sigma_{r\theta} = \left(\sigma_y - \sigma_x\right)\cos\theta\sin\theta + \sigma_{xy}\left(\cos^2\theta - \sin^2\theta\right) \tag{4.51}$$

Applying the definition of the Airy stress function and Eq. 4.38,

$$
\begin{aligned}
\sigma_x &= \frac{\partial^2 \phi}{\partial y^2} = \frac{\partial}{\partial y}\left(\frac{\partial \phi}{\partial y}\right) \\
&= \left(\sin\theta \frac{\partial}{\partial r} + \frac{\cos\theta}{r}\frac{\partial}{\partial \theta}\right)\left(\sin\theta \frac{\partial \phi}{\partial r} + \frac{\cos\theta}{r}\frac{\partial \phi}{\partial \theta}\right) \\
&= \sin^2\theta \frac{\partial^2 \phi}{\partial r^2} + 2\sin\theta\cos\theta \frac{\partial}{\partial r}\left(\frac{1}{r}\frac{\partial \phi}{\partial \theta}\right) + \cos^2\theta\left(\frac{1}{r^2}\frac{\partial^2 \phi}{\partial \theta^2} + \frac{1}{r}\frac{\partial \phi}{\partial r}\right) \\
\sigma_y &= \frac{\partial^2 \phi}{\partial x^2} = \cos^2\theta \frac{\partial^2 \phi}{\partial r^2} - 2\sin\theta\cos\theta \frac{\partial}{\partial r}\left(\frac{1}{r}\frac{\partial \phi}{\partial \theta}\right) + \sin^2\theta\left(\frac{1}{r^2}\frac{\partial^2 \phi}{\partial \theta^2} + \frac{1}{r}\frac{\partial \phi}{\partial r}\right) \\
\sigma_{xy} &= -\frac{\partial^2 \phi}{\partial x \partial y} = \sin\theta\cos\theta\left(\frac{1}{r}\frac{\partial \phi}{\partial r} + \frac{1}{r^2}\frac{\partial^2 \phi}{\partial \theta^2} - \frac{\partial^2 \phi}{\partial r^2}\right) - \left(\cos^2\theta - \sin^2\theta\right)\frac{\partial}{\partial r}\left(\frac{1}{r}\frac{\partial \phi}{\partial \theta}\right)
\end{aligned}
$$

Substituting the results into Eq. 4.51,

$$
\begin{aligned}
\sigma_r &= \frac{1}{r}\frac{\partial \phi}{\partial r} + \frac{1}{r^2}\frac{\partial^2 \phi}{\partial \theta^2} \\
\sigma_\theta &= \frac{\partial^2 \phi}{\partial r^2} \\
\sigma_{r\theta} &= -\frac{\partial}{\partial r}\left(\frac{1}{r}\frac{\partial \phi}{\partial \theta}\right)
\end{aligned}
\tag{4.52}
$$

## 4.4 Axisymmetric Problem

In an axisymmetric problem, the stress, strain, and displacement components do not depend on $\theta$.

***Example 4.9*** Define the Airy stress function for this problem, and determine the stresses when a circular plate with unit thickness is rotated about its center with an angular speed $\omega$. (Reference: A. C. Ugural and S. K. Fenster, Advanced Strength and Applied Elasticity, Elsevier North Holland Publishing Co., Inc., pp. 257–260, 1978).

**Solution**

Consider the element $r\,\mathrm{d}r\,\mathrm{d}\theta$ within the circular plate as shown in Fig. 4.4

In this problem, the centrifugal force $f_r$ is given by,

$$
\begin{aligned}
f_r &= m\frac{v^2}{r} \\
v &= r\omega
\end{aligned}
\tag{4.53}
$$

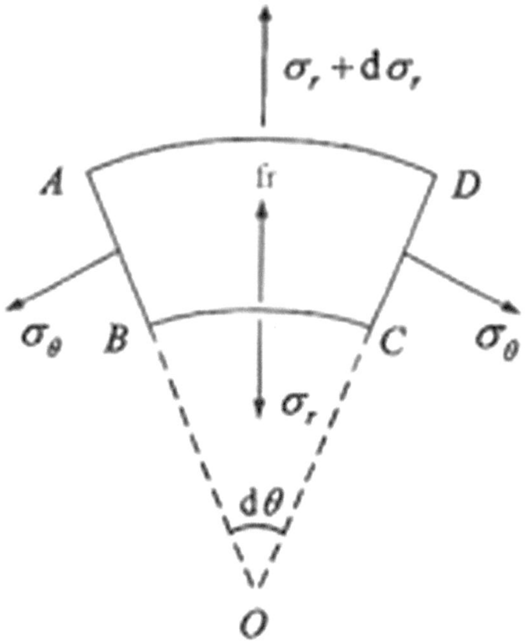


**Fig. 4.4** Element in circular plate

where $m$ is mass, $v$ is the velocity, and $\omega$ is angular velocity.

The centrifugal force per unit volume, $X_r$ is given by Eq. 4.53 as follows,

$$X_r = \rho\omega^2 r \tag{4.54}$$

where $\rho$ is the density of the plate.

Because this problem is an axisymmetric problem, only force equilibrium in the radial direction is considered as follows,

$$\Sigma F_r = 0 \tag{4.55}$$

Using the body force diagram shown in Fig. 4.4,

$$\begin{aligned}(\sigma_r + \mathrm{d}\sigma_r)(r + \mathrm{d}r)\mathrm{d}\theta - \sigma_r r\mathrm{d}\theta \\ - (\sigma_\theta \mathrm{d}r)\left(\sin\frac{\mathrm{d}\theta}{2}\right)2 + X_r(r\mathrm{d}\theta)\mathrm{d}r = 0\end{aligned} \tag{4.56}$$

By neglecting the term with $\mathrm{d}\sigma_r \mathrm{d}r\mathrm{d}\theta$ and using $\sin\theta \approx \theta$ since $\theta$ is small, Eq. 4.56 is written as follows,

$$(\sigma_r - \sigma_\theta)\mathrm{d}r\mathrm{d}\theta + r\mathrm{d}\sigma_r\mathrm{d}\theta + X_r r\mathrm{d}r\mathrm{d}\theta = 0 \tag{4.57}$$

By dividing each term in Eq. 4.57 with $r\,\mathrm{d}r\,\mathrm{d}\theta$

$$\frac{\mathrm{d}\sigma_r}{\mathrm{d}r} + \frac{\sigma_r - \sigma_\theta}{r} + X_r = 0 \tag{4.58}$$

By introducing Eq. 4.54 into Eq. 4.58,

$$\frac{\mathrm{d}\sigma_r}{\mathrm{d}r} + \frac{\sigma_r - \sigma_\theta}{r} + \rho\omega^2 r = 0 \tag{4.59}$$

Equation 4.59 is the equilibrium equation in this problem.
Let us determine the Airy stress function that automatically satisfies Eq. 4.59.
Using Eqs. 4.52 and 4.1, which state,

$$\begin{aligned} \sigma_r &= \frac{1}{r}\frac{\partial \phi}{\partial r} + \frac{1}{r^2}\frac{\partial^2 \phi}{\partial \theta^2} \\ \sigma_\theta &= \frac{\partial^2 \phi}{\partial r^2} \\ \sigma_{r\theta} &= -\frac{\partial}{\partial r}\left(\frac{1}{r}\frac{\partial \phi}{\partial \theta}\right) \end{aligned} \tag{4.60}$$

$$\begin{aligned} \sigma_x + \Omega &= \frac{\partial^2 \phi}{\partial y^2}, \quad \sigma_y + \Omega = \frac{\partial^2 \phi}{\partial x^2} \\ \sigma_{xy} &= -\frac{\partial^2 \phi}{\partial x \partial y} \end{aligned} \tag{4.61}$$

And the fact that stresses should not be dependent on $\theta$, Eq. 4.60 can be expressed as follows:

$$\begin{aligned} \sigma_r &= \frac{1}{r} \cdot \frac{\mathrm{d}\phi}{\mathrm{d}r} + f(r) \\ \sigma_\theta &= \frac{d^2\phi}{\mathrm{d}r^2} + g(r) \\ \sigma_{r\theta} &= 0 \end{aligned} \tag{4.62}$$

By introducing Eq. 4.62 into Eq. 4.59, we can find,

$$\frac{\mathrm{d}f(r)}{\mathrm{d}r} + \frac{f(r)}{r} - \frac{g(r)}{r} + \rho\omega^2 r = 0 \tag{4.63}$$

To satisfy Eq. 4.63, we assume Eq. 4.64.

$$f(r) = 0, \quad g(r) = \rho\omega^2 r^2 \tag{4.64}$$

By introducing Eq. 4.64 into Eq. 4.62,

$$\sigma_r = \frac{1}{r}\frac{\mathrm{d}\phi}{\mathrm{d}r}$$
$$\sigma_\theta = \frac{d}{\mathrm{d}r}\left(\frac{\mathrm{d}\phi}{\mathrm{d}r}\right) + \rho\omega^2 r^2$$
$$\sigma_{r\theta} = 0 \tag{4.65}$$

This is Airy stress function which automatically satisfies equilibrium equation, Eq. 4.59.

The relationship between strain and displacement in polar coordinates, given by Eq. 4.45, is as follows,

$$\varepsilon_r = \frac{\partial u_r}{\partial r}$$
$$\varepsilon_\theta = \frac{u_r}{r} + \frac{1}{r}\frac{\partial u_\theta}{\partial \theta}$$
$$\varepsilon_{r\theta} = \frac{1}{2}\left(\frac{1}{r}\frac{\partial u_r}{\partial \theta} + \frac{\partial u_\theta}{\partial r} - \frac{u_\theta}{r}\right) \tag{4.66}$$

Since this problem is axisymmetric, Eq. 4.66 can be written as follows,

$$\varepsilon_r = \frac{\mathrm{d}u_r}{\mathrm{d}r}$$
$$\varepsilon_\theta = \frac{u_r}{r}$$
$$\varepsilon_{r\theta} = 0 \tag{4.67}$$

Since $\varepsilon_{r\theta} = 0$, the last one of Eq. 4.66 gives

$$\frac{\mathrm{d}u_\theta}{\mathrm{d}r} - \frac{u_\theta}{r} = 0 \tag{4.68}$$

Equation 4.68 means $u_\theta = 0$. We conclude that in an axisymmetric problem, $u_\theta$ should be zero.

From Eq. 4.67,

$$\varepsilon_r = \frac{\mathrm{d}u_r}{\mathrm{d}r} = \frac{\mathrm{d}}{\mathrm{d}r}(r\varepsilon_\theta)$$

Therefore,

$$\frac{\mathrm{d}}{\mathrm{d}r}(r\varepsilon_\theta) - \varepsilon_r = 0 \tag{4.69}$$

This is a compatibility equation for this problem.
To obtain Airy stress function $\phi$, express Eq. 4.69 in terms of $\phi$.
Hooke's Law in plane stress state is as follows,

$$\begin{aligned}\varepsilon_r &= \frac{1}{E}(\sigma_r - \nu\sigma_\theta)\\ \varepsilon_\theta &= \frac{1}{E}(\sigma_\theta - \nu\sigma_r)\end{aligned} \tag{4.70}$$

By introducing Eq. 4.70 into Eq. 4.69,

$$(1+\nu)(\sigma_\theta - \sigma_r) + r\frac{\mathrm{d}\sigma_\theta}{\mathrm{d}r} - \nu r\frac{\mathrm{d}\sigma_r}{\mathrm{d}r} = 0 \tag{4.71}$$

By introducing Eq. 4.65 into Eq. 4.71,

$$\frac{d^2}{\mathrm{d}r^2}\frac{\mathrm{d}\phi}{\mathrm{d}r} + \frac{1}{r}\frac{d}{\mathrm{d}r}\left(\frac{\mathrm{d}\phi}{\mathrm{d}r}\right) - \frac{1}{r^2}\frac{\mathrm{d}\phi}{\mathrm{d}r} + (3+\nu)\rho\omega^2 r = 0 \tag{4.72}$$

By letting $F = \frac{\mathrm{d}\phi}{\mathrm{d}r}$, the first three terms of Eq. 4.72 are written as follows,

$$\begin{aligned}&\frac{d^2}{\mathrm{d}r^2}\left(\frac{\mathrm{d}\phi}{\mathrm{d}r}\right) + \frac{1}{r}\frac{d}{\mathrm{d}r}\left(\frac{\mathrm{d}\phi}{\mathrm{d}r}\right) - \frac{1}{r^2}\frac{\mathrm{d}\phi}{\mathrm{d}r}\\ &= \frac{\mathrm{d}^2F}{\mathrm{d}r^2} + \frac{1}{r}\frac{\mathrm{d}F}{\mathrm{d}r} - \frac{F}{r^2}\\ &= \frac{\mathrm{d}}{\mathrm{d}r}\left(\frac{\mathrm{d}F}{\mathrm{d}r}\right) + \frac{\mathrm{d}}{\mathrm{d}r}\left(\frac{F}{r}\right)\\ &= \frac{\mathrm{d}}{\mathrm{d}r}\left[\frac{\mathrm{d}F}{\mathrm{d}r} + \frac{F}{r}\right]\\ &= \frac{\mathrm{d}}{\mathrm{d}r}\left[\frac{1}{r}\left(r\frac{\mathrm{d}F}{\mathrm{d}r} + F\right)\right]\\ &= \frac{\mathrm{d}}{\mathrm{d}r}\left[\frac{1}{r}\frac{\mathrm{d}}{\mathrm{d}r}(rF)\right]\end{aligned}$$

Therefore, Eq. 4.72 becomes,

$$\frac{d}{\mathrm{d}r}\left[\frac{1}{r}\frac{d}{\mathrm{d}r}(rF)\right] = -(3+\nu)\rho\omega^2 r \tag{4.73}$$

By introducing $F = \frac{\mathrm{d}\phi}{\mathrm{d}r}$ into Eq. 4.73,

$$\frac{d}{\mathrm{d}r}\left[\frac{1}{r}\frac{d}{\mathrm{d}r}\left(r\frac{\mathrm{d}\phi}{\mathrm{d}r}\right)\right] = -(3+\nu)\rho\omega^2 r \tag{4.74}$$

By integrating Eq. 4.74,

$$\frac{\mathrm{d}\phi}{\mathrm{d}r} = -\frac{3+\nu}{8}\rho\omega^2 r^3 + C_1\frac{r}{2} + C_2\frac{1}{r} \tag{4.75}$$

where $C_1$ and $C_2$ are real constant as integral constants.

By introducing Eq. 4.75 into Eq. 4.65,

$$\begin{aligned}\sigma_r &= -\frac{3+\nu}{8}\rho\omega^2 r^2 + \frac{C_1}{2} + \frac{C_2}{r^2}\\ \sigma_\theta &= -\frac{1+3\nu}{8}\rho\omega^2 r^2 + \frac{C_1}{2} - \frac{C_2}{r^2}\\ \sigma_{r\theta} &= 0\end{aligned} \tag{4.76}$$

1. **In case of a circular plate with a radius R**

When there is no rotation, since stresses at the center of circular plate should not be infinite,

$$C_2 = 0 \tag{4.77}$$

Since $\sigma_r$ should be zero at $r = R$,

$$C_1 = \frac{3+r}{4}\rho\omega^2 R^2 \tag{4.78}$$

Therefore, by introducing Eqs. 4.77 and 4.78 into Eq. 4.76,

$$\begin{aligned}\sigma_r &= \frac{3+\nu}{8}\rho\omega^2(R^2 - r^2)\\ \sigma_\theta &= \frac{\rho\omega^2}{8}\left[(3+\nu)R^2 - (1+3\nu)r^2\right]\end{aligned} \tag{4.79}$$

According to Eq. 4.79, maximum stress occurs at $r = 0$.

$$(\sigma_r)_{r=0} = (\sigma_\theta)_{r=0} = \frac{3+\nu}{8}\rho\omega^2 R^2 \tag{4.80}$$

2. **In case of a circular plate with an inner hole radius $R_1$ and outer radius $R_2$**

Since $\sigma_r = 0$ at $r = R_1$, from Eq. 4.76,

$$(\sigma_r)_{r=R_1} = -\frac{3+\nu}{8}\rho\omega^2 R_1^2 + \frac{C_1}{2} + \frac{C_2}{R_1^2} = 0 \tag{4.81}$$

Since $\sigma_r = 0$ at $r = R_2$, from Eq. 4.76,

$$(\sigma_r)_{r=R_2} = -\frac{3+\nu}{8}\rho\omega^2 R_2^2 + \frac{C_1}{2} + \frac{C_2}{R_2^2} = 0 \tag{4.82}$$

Get $C_1$ and $C_2$ from Eqs. 4.81 and 4.82 and introduce the results into Eq. 4.76. Finally, we get,

$$\begin{aligned}\sigma_r &= \frac{3+\nu}{8}\rho\omega^2\left(R_1^2 + R_2^2 - \frac{R_1^2R_2^2}{r^2} - r^2\right)\\ \sigma_\theta &= \frac{3+\nu}{8}\rho\omega^2\left(R_1^2 + R_2^2 + \frac{R_1^2R_2^2}{r^2} - \frac{1+3\nu}{3+\nu}r^2\right)\end{aligned} \tag{4.83}$$

According to Eq. 4.83, maximum stress occurs at $(\sigma_\theta)_{r=R_1}$ as follows,

$$(\sigma_\theta)_{r=R_1} = (\sigma_\theta)_{\max} = \frac{3+\nu}{4}\rho\omega^2 R_2^2\left(1 + \frac{1-\nu}{3+\nu}\frac{R_1^2}{R_2^2}\right) \tag{4.84}$$

When $\frac{R_1}{R_2}$ approaches to zero,

$$(\sigma_\theta)_{\max} = \frac{3+\nu}{4}\rho\omega^2 R_2^2 \tag{4.85}$$

Compare Eq. 4.85 with Eq. 4.80 to observe the difference in maximum stress between the two cases. It is noted that if a crack is initiated, it can be initiated at the center in case that there is no hole at the center and at the inner circle of the hole in case that there is a hole at the center, as shown in the example.

If a crack is initiated, it will start at the center if there is no hole, or at the inner edge of the hole if one is present, as shown in the example. Additionally, if a crack is initiated, it should propagate in the radial direction, since the maximum stress occurs in the term $\sigma_\theta$ as shown in the example.

# Part II
# Fracture

# Chapter 5
# Stress Intensity Factor

**Abstract** The difference between notch and crack, the concept, physical meaning, and importance of stress intensity factor, and theoretical and experimental plastic zone shapes and sizes at the crack tip are explained.

## 5.1 Maximum Theoretical Failure Strength

It is assumed that a material has no defects. Figure 5.1 illustrates the elongation caused by tensile stress applied to the atom.

Assuming the material is perfect, the critical stress $\sigma_c$ can be determined using linear elasticity theory as follows,

$$\sigma_c = \frac{E\lambda}{2\pi a_0} \tag{5.1}$$

where $E$ is Young's modulus, $a_0$ is the atomic distance before deformation, and $\lambda$ is the wavelength of the stress curve. Since $a_0 \approx \lambda$ in order, $\sigma_c \approx \frac{E}{2\pi} \approx \frac{E}{10}$ in order.

Therefore,

$$\frac{E}{\sigma_c} \approx 10 \tag{5.2}$$

Table 5.1 shows a discrepancy between theoretical and experimental results, which arises from the incorrect assumption that the material has no defects.

## 5.2 Griffith Model

When Griffith developed his theory, he assumed that a material might not be perfect and could contain a crack, as shown in Fig. 5.2. This assumption has been confirmed to be true today.

 
K. Y. Lee, *Introduction to Elasticity, Fracture and Fatigue*,
https://doi.org/10.1007/978-981-96-3600-6_5

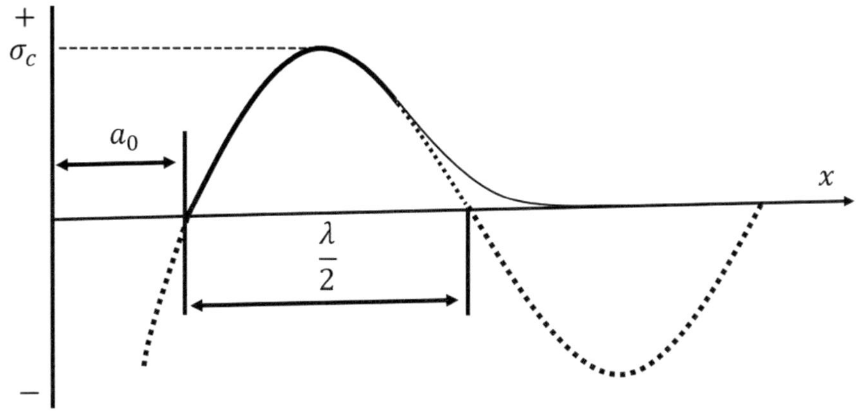


**Fig. 5.1** Stress versus elongation

**Table 5.1** Experimental data of Young's modulus versus critical stress

| Material | $E/\sigma_c$ |
|---|---|
| Iron whisker | 23 |
| Silicon whisker | 26 |
| Alumina whisker | 33 |
| Ausformed steel | 64 |
| Piano wire | 73 |

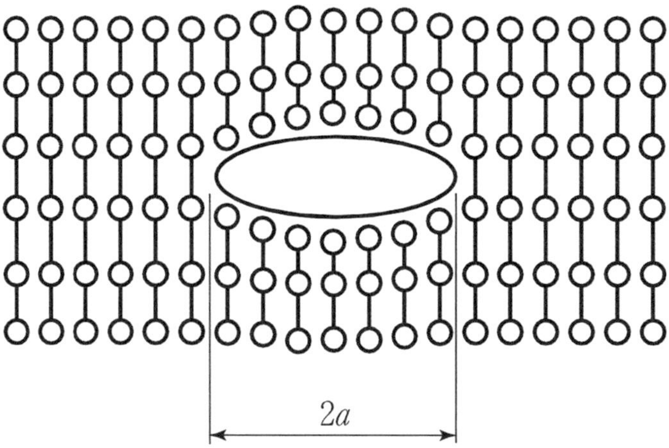


**Fig. 5.2** Crack in the material

Griffith derived Eq. 5.3 using linear elasticity theory.

$$\sigma_c = \left(\frac{2E\gamma_s}{\pi a}\right)^{\frac{1}{2}} \tag{5.3}$$

where $\gamma_s$ is the surface energy per unit area and $a$ is the half-length of a crack. $\gamma_s$ represents the energy per unit area required to produce one side of the crack surface. In this book, micro-cracks are not considered. Whenever we refer to a crack, it always denotes a macro-crack, which is defined as larger than the grain size. $\sigma_c$ in Eq. 5.3 is much smaller than in Eq. 5.1 since $a$ is much greater than $a_0$.

He demonstrated that Eq. 5.3 agrees with experimental results by testing glass, a linear elastic material. By publishing one theoretical and one experimental research manuscript related to Eq. 5.3 and proving that the theory and experiment agreed with each other, he became a pioneer in fracture mechanics.

## 5.3 Difference Between Notch and Crack

Figure 5.3 shows an elliptical notch in an infinite elastic body under applied stress.

Using linear elasticity theory, the stress concentration factor $K_t$ can be derived as follows,

$$K_t = 1 + 2\sqrt{\frac{a}{\rho}} \tag{5.4}$$

**Fig. 5.3** Elliptical notch in infinite body

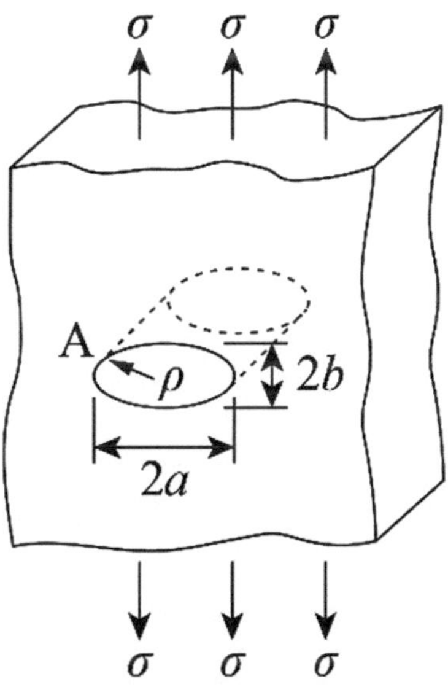

where $\rho$ is the radius of notch root and $a$ is the half-length of notch.

Note that $K_t$ depends only on the shape and size of the notch, as the body is infinite. By definition, the stress concentration factor cannot depend on stress or load. However, if the body is finite, $K_t$ would depend on the shape and size of the body.

Since $\rho$ always has a finite value in case of a notch,$K_t$ also has a finite value. But in case of $\rho \rightarrow 0$, $K_t \rightarrow \infty$. This case is called a crack.

Since $K_t$ for a crack always has an infinite value, we can no longer use $K_t$ as a parameter for cracks. Note that a crack is much more dangerous than a notch, as $K_t$ is always infinite for a crack but finite for a notch.

***Example 5.1*** If a crack is found at the edge of a large cast iron bell, what measures can be taken to prevent fracture without welding the crack?

**Solution**

It is recommended to turn the crack into a notch to reduce stress at the crack tip. However, it is crucial to accurately locate the crack tip before converting it into a notch. Creating a notch near, but not precisely at, the crack tip can exacerbate the stress state at the crack tip due to the presence of the notch.

## 5.4 Stress Intensity Factor

Figure 5.4 illustrates three types of crack deformation: Mode I, known as the opening mode; Mode II, the sliding mode; and Mode III, the tearing mode.

***Example 5.2*** Which mode in Fig. 5.4 is considered the most dangerous?

**Solution**

Imagine your mouth as a crack and let's say you apply equal force to that mouth in three different modes. You will find that the most painful mode is Mode I. Since Mode I is the most dangerous, it is emphasized throughout this book. Figure 5.5

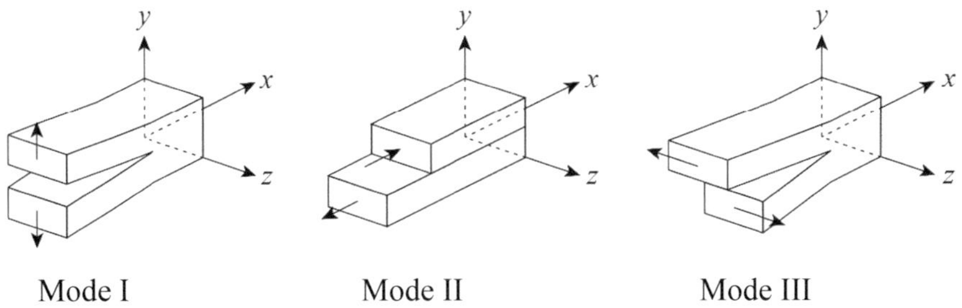


**Fig. 5.4** Three modes of crack deformation

Fig. 5.5 Crack tip stress components in Mode I

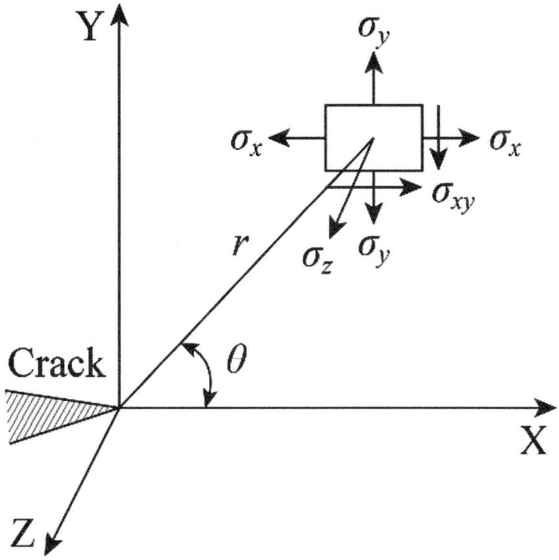


shows the stress components near the crack tip in Mode I, and Eq. 5.5 can be derived using linear elasticity theory.

$$
\begin{aligned}
\sigma_x &= \frac{K_\mathrm{I}}{\sqrt{2\pi r}} \cos\frac{\theta}{2}\left(1 - \sin\frac{\theta}{2}\sin\frac{3\theta}{2}\right) \\
\sigma_y &= \frac{K_\mathrm{I}}{\sqrt{2\pi r}} \cos\frac{\theta}{2}\left(1 + \sin\frac{\theta}{2}\sin\frac{3\theta}{2}\right) \\
\sigma_{xy} &= \frac{K_\mathrm{I}}{\sqrt{2\pi r}} \left(\sin\frac{\theta}{2}\cos\frac{\theta}{2}\cos\frac{3\theta}{2}\right) \qquad (5.5)
\end{aligned}
$$

where $K_\mathrm{I}$ is known as the Mode I stress intensity factor, and $r$ and $\theta$ are polar coordinates.

This equation is referred to as the stress field equation near a crack tip in Mode I within two-dimensional fracture mechanics. Note that additional terms, except the singular term shown in Eq. 5.5, are neglected near the crack tip as they are negligible compared to the singular term. This equation applies to any shape and size of a two-dimensional body.

Characteristics of Eq. 5.5 are as follows,

1. $\sigma_{ij} \propto \frac{1}{\sqrt{r}}$

If $r$ approaches zero, $\sigma_{ij}$ becomes infinite.

This infinity is possible within the theory of linear elasticity but not in reality, a phenomenon known as stress singularity. In practice, $\sigma_{ij}$ cannot become infinite because the material yields before $\sigma_{ij}$ reaches infinity.

2. $\sigma_{ij} \neq f(z)$

Since Eq. 5.5 is a two-dimensional linear elastic solution, $\sigma_{ij}$ cannot be a function of $z$. In this book, the $z$-direction corresponds to the thickness of the body.

3. $\sigma_{ij} = f(r, \theta, K_{\mathrm{I}})$

If you choose the location $(r, \theta)$ near the crack tip, the magnitude of $\sigma_{ij}$ at the coordinate depends only on $K_{\mathrm{I}}$ In other word, $\sigma_{ij} \propto f(K_{\mathrm{I}})$ at $(r, \theta)$.

However, we know that the stress field $\sigma_{ij}$ should depend on the size and shape of both the crack and the body, as well as the applied stress. This implies that $K_{\mathrm{I}}$ should have also the same dependence as stress field $\sigma_{ij}$.

In other word,

$\sigma_{ij}, K_{\mathrm{I}} = f$ (sizes and shapes of a crack and body, and applied stress).

4. If $K_{\mathrm{I}}$ is known, substitute it into Eq. 5.5. The resulting expressions are the stress field equations near the crack tip.

From Eq. 5.5 and Hooke's Law, strain field equations near the crack tip can be derived. Displacement field equations near the crack tip can then be obtained using the relationships between strains and displacements. This means that if $K_{\mathrm{I}}$ is known, the stress, strain, and displacement field equations can be determined using elasticity theory. This is why we focus on finding the stress intensity factor rather than directly determining the stress, strain, and displacement field equations. It is noted that the general form of $K_{\mathrm{I}}$ is as follows,

$$K_{\mathrm{I}} = Y\sigma\sqrt{a} \tag{5.6}$$

where $Y$ is a polynomial function of $\frac{a}{W}$. $W$ is the width of a specimen.

***Example 5.3*** Is it physically acceptable for the magnitude of the stress intensity factor to decrease or have an inflection point as crack length increases?

**Solution**

Since stress near the crack tip should physically increase as $a$ increases under constant applied stress, $K$ should also increase accordingly.

***Example 5.4*** What is the primary functional difference between the stress concentration factor and the stress intensity factor?

**Solution**

The stress intensity factor depends on the applied stress, whereas the stress concentration factor does not.

The Mode II stress field equations are shown as follows,

$$\sigma_x = -\frac{K_{\mathrm{II}}}{\sqrt{2\pi r}} \sin\frac{\theta}{2}\left(2 + \cos\frac{\theta}{2}\cos\frac{3\theta}{2}\right)$$

$$\sigma_y = \frac{K_{II}}{\sqrt{2\pi r}} \sin\frac{\theta}{2}\cos\frac{\theta}{2}\cos\frac{3\theta}{2}$$

$$\sigma_{xy} = \frac{K_{II}}{\sqrt{2\pi r}} \cos\frac{\theta}{2}\left(1 - \sin\frac{\theta}{2}\sin\frac{3\theta}{2}\right) \tag{5.7}$$

The Mode III stress field equations are shown as follows,

$$\sigma_{zx} = -\frac{K_{III}}{\sqrt{2\pi r}} \sin\frac{\theta}{2}$$

$$\sigma_{zy} = \frac{K_{III}}{\sqrt{2\pi r}} \cos\frac{\theta}{2}$$

$$\sigma_x = 0,\ \sigma_y = 0,\ \sigma_z = 0,\ \sigma_{xy} = 0 \tag{5.8}$$

***Example 5.5*** Why must $\sigma_x = 0$, $\sigma_y = 0$, $\sigma_z = 0$, and $\sigma_{xy} = 0$ be satisfied in Eq. 5.8?

**Solution**

Equation 5.8 corresponds to Mode III. In Mode III, only a tearing shear force in the z-direction is applied, so $\sigma_x$, $\sigma_y$, $\sigma_z$, and $\sigma_{xy}$ do not occur.

The general forms of the stress field equations in Modes I, II, and III, according to Eqs. 5.5, 5.7, and 5.8, are as follows,

$$\sigma_{ij} \propto \frac{K}{\sqrt{r}} f_{ij}(\theta) \quad (i, j = x, y) \tag{5.9}$$

where the general stress intensity factor forms in Mode I, II and III are as follows,

$$K = Y\left(\frac{a}{W}\right)\sigma\sqrt{a} \tag{5.10}$$

In Eq. 5.10, for Mode I, $\sigma$ represents the applied normal stress, for Mode II, σ represents the applied sliding shear stress, and for Mode III, $\sigma$ represents the applied tearing shear stress.

## 5.5 Approximation and Superposition Methods of Stress Intensity Factor

The stress intensity factor can be determined using elasticity theory, the finite element method, or the boundary element method. However, these approaches require significant effort and time. Therefore, approximation and superposition methods are introduced to simplify the process of finding the stress intensity factor.

### 5.5.1 Approximation of Stress Intensity Factor

Some examples are provided below:

1. If there is a crack around a hole and the crack length is very small compared to the hole, it is possible to assume the crack as an edge crack under a concentrated load near the hole.
2. If there is a crack around a very small hole but the crack length is very large compared to the hole, it is possible to assume the crack as a center crack, where the crack length is the sum of the hole diameter and the crack length.
3. If the body is large enough compared to the crack, even if the body is finite, it is possible to approximate the body as an infinite body.

### 5.5.2 Superposition of Stress Intensity Factor

Superposition relates to the applied force or stress. Stress can be superposed because it is linearly related to force. Similarly, the stress intensity factor can be superposed, as it is linearly related to stress, as shown in Eq. 5.10.

***Example 5.6*** Can strain energy density be superposed?

**Solution**

Strain energy density $w$ in tension test is written as follows,

$$w = \frac{1}{2}\sigma\varepsilon \tag{5.11}$$

where $\sigma$ is stress and $\varepsilon$ strain.

$$\therefore w = \frac{1}{2}\sigma\varepsilon = \frac{1}{2}\sigma\frac{\sigma}{E} = \frac{\sigma^2}{2E}$$

$$\text{If } \sigma = \sigma_1, w_1 = \frac{\sigma_1^2}{2E}$$

$$\text{If } \sigma = \sigma_2, w_2 = \frac{\sigma_2^2}{2E}$$

$$\therefore \text{If } \sigma = \sigma_1 + \sigma_2, w = \frac{(\sigma_1 + \sigma_2)^2}{2E}$$

$$= \frac{\sigma_1^2}{2E} + \frac{\sigma_2^2}{2E} + \frac{2\sigma_1\sigma_2}{2E} = w_1 + w_2 + \frac{\sigma_1\sigma_2}{E}$$

Therefore, strain energy density cannot be superposed with respect to applied stress or force as it is not linear with respect to stress and include a coupling term $\frac{\sigma_1\sigma_2}{E}$ in addition to $w_1 + w_2$.

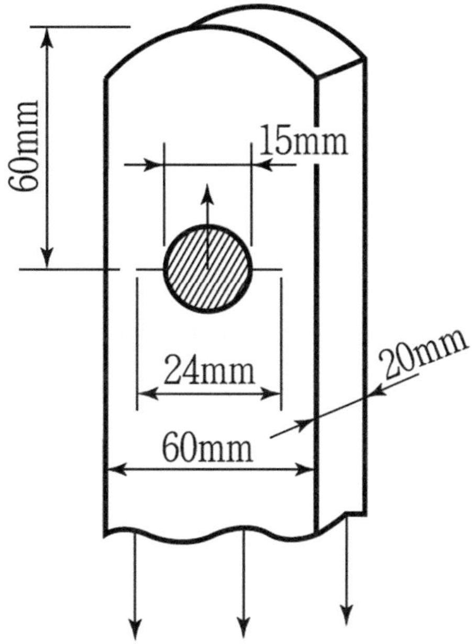


**Fig. 5.6** Crack in the pin-loaded lug

***Example 5.7*** Consider a crack in the pin-loaded lug used in the bridge as shown in Fig. 5.6. Determine the stress intensity factor for this model using the superposition method of stress intensity factors. (Reference: A.P. Parker, *The Mechanics of Fracture and Fatigue: An Introduction,* E. & F. N. Spon LTD, London, pp. 157–159, 1981).

**Solution**

First, for simplicity, the lug is assumed to be infinitely long.

Based on the superposition of stress intensity factors, we can construct Fig. 5.7.

Figure 5.7c represents the model for which we want to determine the stress intensity factor.

$$\therefore K = \frac{1}{2}(K_P + K_\sigma) \tag{a}$$

The solid lines are derived from the stress intensity factor handbook. The dotted line in Fig. 5.8 represents the solution according to Equation a.

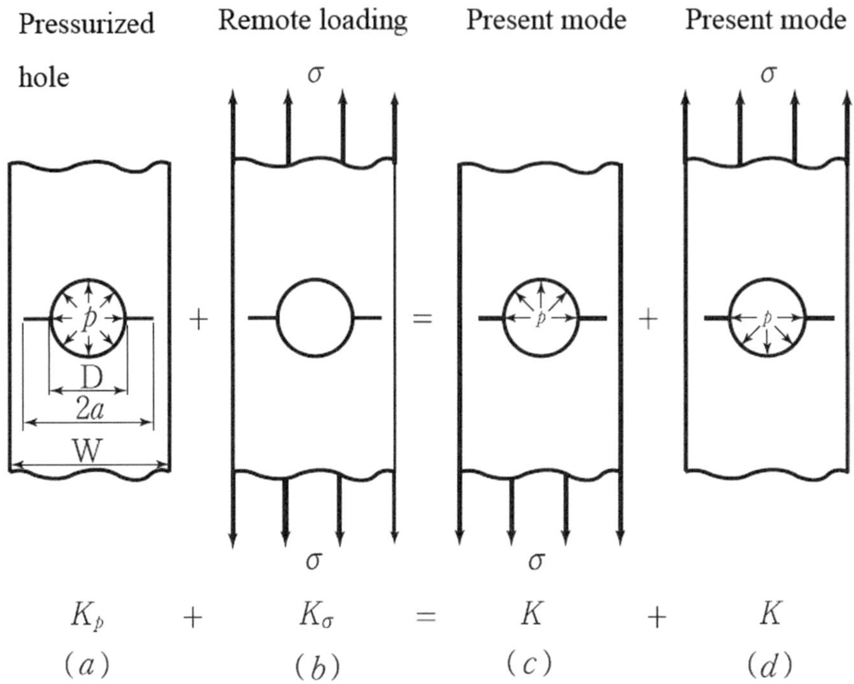


**Fig. 5.7** Superposition of stress intensity factors

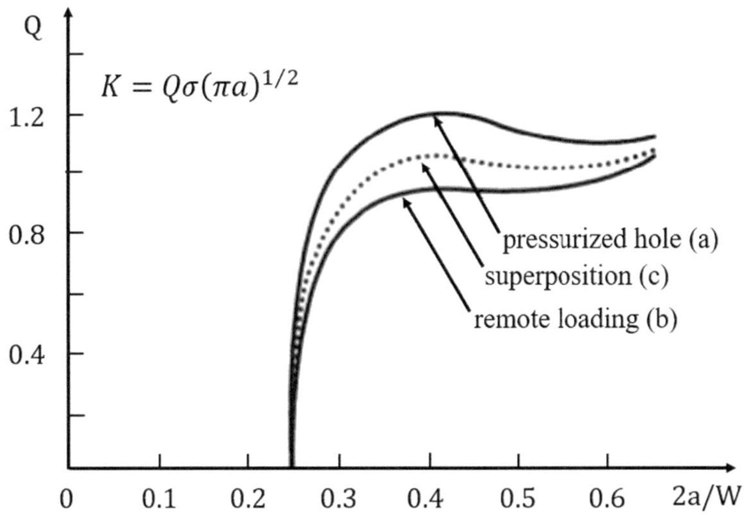


**Fig. 5.8** Stress intensity factor for a crack in the pin-loaded lug

## 5.6 Plastic Zone Size and Shape at Crack Tip

Since theoretical stresses induced at the crack tip always have infinite values, as shown in Eqs. 5.5, 5.7, and 5.8, plastic zones are expected to form at the crack tip in real materials, even under very low applied stress. Irwin's empirical plastic zone sizes in the plane stress and plane strain states are as follows,

$$r_p = \begin{cases} \frac{1}{\pi}\left(\frac{K_I}{\sigma_{ys}}\right)^2 & \text{(Plane stress state)} \\ \frac{1}{3\pi}\left(\frac{K_I}{\sigma_{ys}}\right)^2 & \text{(Plane strain state)} \end{cases} \tag{5.12}$$

From Eq. 5.12, it is noted that the plastic zone size at the crack tip is larger in the plane stress state than in the plane strain state.

The theoretical sizes and shapes of plastic zone at the crack tip in the plane stress and strain states can be obtained by applying yield criteria.

### *5.6.1 Theoretical Sizes and Shapes of Plastic Zone at the Crack Tip in the Plane Stress State*

In the plane stress state, stress field equations can be shown as follows,

$$\begin{aligned}
\sigma_x &= \frac{K_I}{\sqrt{2\pi r}}\cos\frac{\theta}{2}\left(1 - \sin\frac{\theta}{2}\sin\frac{3\theta}{2}\right) \\
\sigma_y &= \frac{K_I}{\sqrt{2\pi r}}\cos\frac{\theta}{2}\left(1 + \sin\frac{\theta}{2}\sin\frac{3\theta}{2}\right) \\
\sigma_{xy} &= \frac{K_I}{\sqrt{2\pi r}}\left(\sin\frac{\theta}{2}\cos\frac{\theta}{2}\cos\frac{3\theta}{2}\right) \\
\sigma_z &= 0,\ \sigma_{zx} = 0,\ \sigma_{zy} = 0
\end{aligned} \tag{5.13}$$

Note that $\sigma_z = 0, \sigma_{zx} = 0, \sigma_{zy} = 0$ in Eq. 5.13 due to the definition of the plane stress state.

The principal stresses are obtained as follows,

$$\begin{aligned}
\sigma_1 &= \frac{K_I}{\sqrt{2\pi r}}\cos\frac{\theta}{2}\left(1 + \sin\frac{\theta}{2}\right) \\
\sigma_2 &= \frac{K_I}{\sqrt{2\pi r}}\cos\frac{\theta}{2}\left(1 - \sin\frac{\theta}{2}\right) \\
\sigma_3 &= 0
\end{aligned} \tag{5.14}$$

where $\sigma_1 > \sigma_2 > \sigma_3$. $\sigma_1$, $\sigma_2$, and $\sigma_3$ are principal stresses.

In Eq. 5.14, $\sigma_1$ and $\sigma_2$ can be obtained from $\sigma_x$, $\sigma_y$, and $\sigma_{xy}$ given in Eq. 5.13 using Mohr's stress circle. $\sigma_3 = 0$ follows directly from $\sigma_z = 0, \sigma_{zx} = 0, \sigma_{zy} = 0$.

1. **Plastic zone size and shape according to Tresca yield criterion**

   Tresca yield criterion is expressed as follows,

$$\sigma_1 - \sigma_3 = \sigma_{ys} \tag{5.15}$$

where $\sigma_{ys}$ is yield stress.

By introducing Eqs. 5.14 and 5.15,

$$r_{\mathrm{p}} = \frac{1}{2\pi}\left(\frac{K_{\mathrm{I}}}{\sigma_{\mathrm{ys}}}\right)^2 \cos^2\frac{\theta}{2}\left(1 + \sin\frac{\theta}{2}\right)^2 \tag{5.16}$$

where $r_{\mathrm{p}}$ represents the size and shape of plastic zone size.

2. **Plastic zone size and shape according to von Mises yield criterion**

   von Mises yield criterion is expressed as follows,

$$(\sigma_1 - \sigma_2)^2 + (\sigma_2 - \sigma_3)^2 + (\sigma_3 - \sigma_1)^2 = 2\sigma_{\mathrm{ys}}^2 \tag{5.17}$$

By introducing Eqs. 5.14 to 5.17,

$$r_{\mathrm{p}} = \frac{1}{4\pi}\left(\frac{K_{\mathrm{I}}}{\sigma_{\mathrm{ys}}}\right)^2 \left(1 + \cos\theta + \frac{3}{2}\sin^2\theta\right) \tag{5.18}$$

### 5.6.2 *Theoretical Sizes and Shapes of the Plastic Zone at the Crack Tip in the Plane Strain State*

In the plane strain state, the stress field equations near the crack tip are expressed as follows,

$$\begin{aligned}
\sigma_x &= \frac{K_{\mathrm{I}}}{\sqrt{2\pi r}}\cos\frac{\theta}{2}\left(1 - \sin\frac{\theta}{2}\sin\frac{3\theta}{2}\right)\\
\sigma_y &= \frac{K_{\mathrm{I}}}{\sqrt{2\pi r}}\cos\frac{\theta}{2}\left(1 + \sin\frac{\theta}{2}\sin\frac{3\theta}{2}\right)\\
\sigma_{xy} &= \frac{K_{\mathrm{I}}}{\sqrt{2\pi r}}\left(\sin\frac{\theta}{2}\cos\frac{\theta}{2}\cos\frac{3\theta}{2}\right)\\
\sigma_z &= \nu(\sigma_x + \sigma_y) = \frac{2\nu K_{\mathrm{I}}}{\sqrt{2\pi r}}\cos\frac{\theta}{2},\ \sigma_{zx} = 0,\ \sigma_{zy} = 0
\end{aligned} \tag{5.19}$$

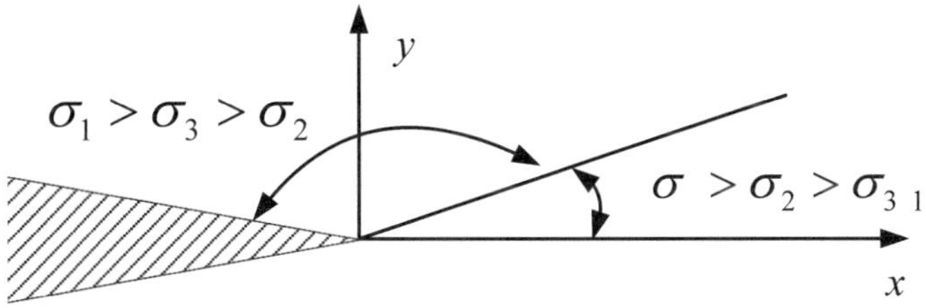


**Fig. 5.9** Magnitudes of $\sigma_1$, $\sigma_2$, and $\sigma_3$

Note that $\sigma_z = \nu(\sigma_x + \sigma_y) = \frac{2\nu K_I}{\sqrt{2\pi r}} \cos\frac{\theta}{2}$, $\sigma_{zx} = 0$, $\sigma_{zy} = 0$ in Eq. 5.19. These expressions can be derived from the definition of the plane strain state, generalized Hooke's law and the fact that $\sigma_z$ becomes principal stress since $\sigma_{zx} = 0$ and $\sigma_{zy} = 0$. Using Mohr's stress circle and Eq. 5.19,

$$\begin{aligned}\sigma_1 &= \frac{K_I}{\sqrt{2\pi r}} \cos\frac{\theta}{2}\left(1 + \sin\frac{\theta}{2}\right) \\ \sigma_2 &= \frac{K_I}{\sqrt{2\pi r}} \cos\frac{\theta}{2}\left(1 - \sin\frac{\theta}{2}\right) \\ \sigma_3 &= \nu(\sigma_1 + \sigma_2) = 2\nu\frac{K_I}{\sqrt{2\pi r}} \cos\frac{\theta}{2}\end{aligned} \tag{5.20}$$

By comparing the magnitude of $\sigma_1$, $\sigma_2$, and $\sigma_3$ across different values of $\theta$, we can observe $\sigma_1 > \sigma_2 > \sigma_3$ for certain $\theta$ values, while $\sigma_1 > \sigma_3 > \sigma_2$ for other $\theta$ values, as illustrated in Fig. 5.9.

1. **Plastic zone size and shape according to Tresca yield criterion**

In case of $\sigma_1 > \sigma_2 > \sigma_3$, by substituting Eq. 5.20 into Eq. 5.15,

$$r_p = \frac{1}{2\pi}\left(\frac{K_I}{\sigma_{ys}}\right)^2 \cos^2\frac{\theta}{2}\left(1 - 2\nu + \sin\frac{\theta}{2}\right)^2 \tag{5.21}$$

In case of $\sigma_1 > \sigma_3 > \sigma_2$, Tresca yield criterion becomes

$$\sigma_1 - \sigma_2 = \sigma_{ys} \tag{5.22}$$

By substituting Eq. 5.20 into Eq. 5.22,

$$r_E = \frac{1}{2\pi}\left(\frac{K_I}{\sigma_{ys}}\right)^2 \sin^2\theta \tag{5.23}$$

2. **Plastic zone size and shape according to von Mises yield criterion**

By substituting Eq. 5.20 into Eq. 5.17,

$$r_{\mathrm{p}} = \frac{1}{4\pi}\left(\frac{K_{\mathrm{I}}}{\sigma_{\mathrm{ys}}}\right)^2\left\{\frac{3}{2}\sin^2\theta + (1-2\nu)^2(1+\cos\theta)\right\} \tag{5.24}$$

Using Eqs. 5.16, 5.18, 5.21, 5.23, and 5.24 we can construct Fig. 5.10.

Note that the theoretical solutions differ from the experimental results, as shown in Fig. 5.11.

However, the experimental and theoretical results agree in the following two respects,

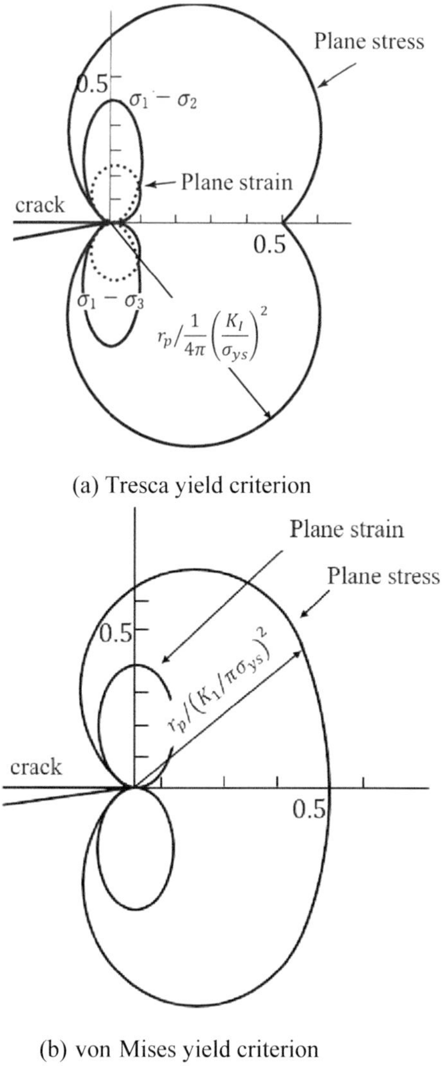


**Fig. 5.10** Theoretical plastic zone shapes and sizes at the crack tip

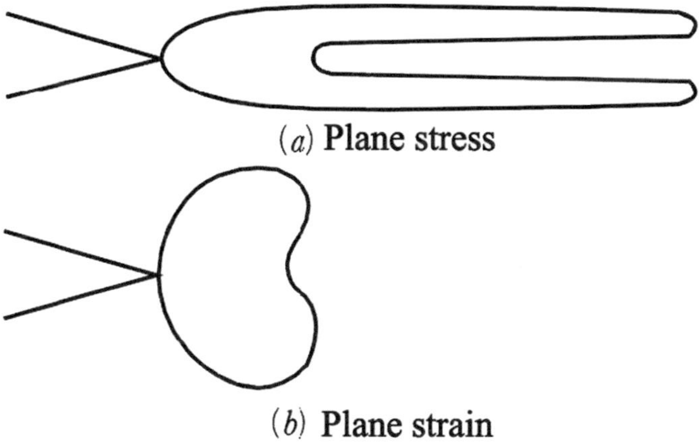


**Fig. 5.11** Experimental plastic zone sizes and shapes at the crack tip

1. $r_{\mathrm{p}} \propto \left(\frac{K_{\mathrm{I}}}{\sigma_{\mathrm{ys}}}\right)^2$ (5.25)

2. The plastic zone size in the plane stress is larger than that in the plane strain.

## 5.7 Relationship Between Theories of Griffith and Irwin

There is a relationship between the theories of Griffith and Irwin as follows,

$$G_{\mathrm{I}} = \begin{cases} \frac{K_{\mathrm{I}}^2}{E} & \text{(Plane stress state)} \\ \frac{K_{\mathrm{I}}^2(1-\nu^2)}{E} & \text{(Plane strain state)} \end{cases} \tag{5.26}$$

where $G_{\mathrm{I}}$ is the Mode I energy release rate obtained from Griffith's energy concept and $K_{\mathrm{I}}$ is the stress intensity factor obtained from Irwin's theory. $E$ and $\nu$ are Young's modulus and Poisson ratio, respectively, which are material constants. Since only material constants relate $G_{\mathrm{I}}$ and $K_{\mathrm{I}}$, the physical meanings of $G_{\mathrm{I}}$ and $K_{\mathrm{I}}$ can be considered equivalent.

## 5.8 Relationship Among CTOD, G, and K

The Crack Tip Opening Displacement (CTOD) is the opening displacement at the tip of a crack caused by an applied load, as shown in Fig. 5.12.

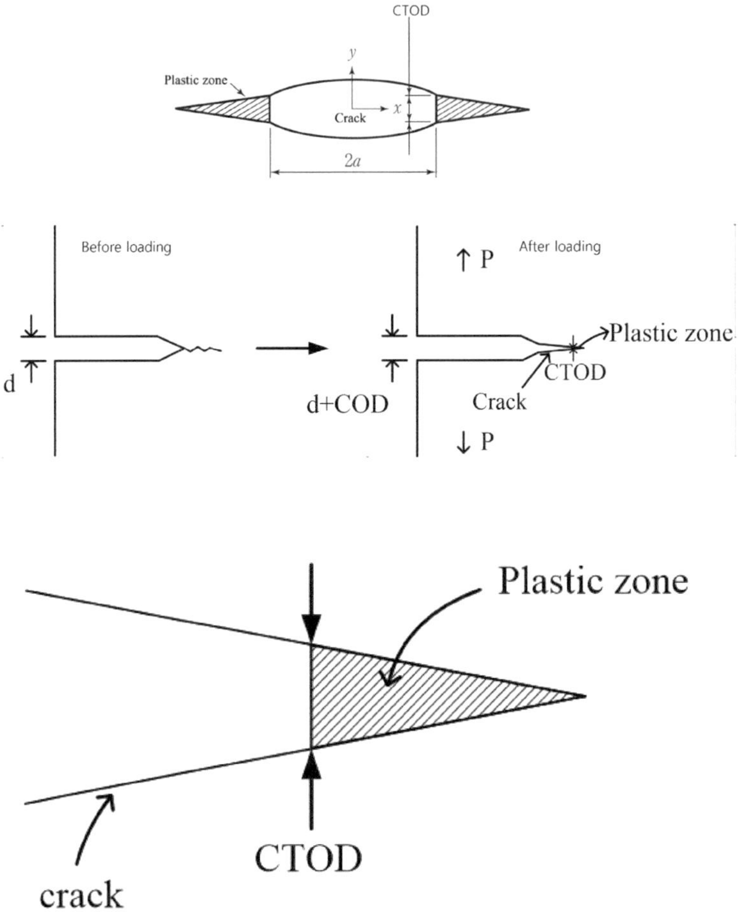


**Fig. 5.12** CTOD and COD

In Fig. 5.12, 2a represents the crack length, *d* is the notch width, and COD (Crack Opening Displacement) is the opening displacement at the mouth of the notch under the applied load.

Equation 5.27 shows the relationship among CTOD, $G_{\mathrm{I}}$, and $K_{\mathrm{I}}$.

$$
\begin{aligned}
&G_{\mathrm{I}} = (\mathrm{CTOD})\sigma_{\mathrm{ys}} = \tfrac{K_{\mathrm{I}}^2}{E} \quad \text{(Plane stress state)} \\
&G_{\mathrm{I}} = n(\mathrm{CTOD})\sigma_{\mathrm{ys}} = \frac{K_{\mathrm{I}}^2(1-\nu^2)}{E} \quad \text{(Plane strain state)} \\
&(1 < n < 2)
\end{aligned} \tag{5.27}
$$

CTOD is measured according to *Methods for Crack Opening Displacement Testing, British Standard Institution BS 5762–1979, 1979.*

# Chapter 6
# Plane Strain Fracture Toughness

**Abstract** The standard test method of plane strain fracture toughness is introduced. The fracture mechanics design method to prevent yielding and fracture is also explained.

## 6.1 Transition of Fracture Surface

The stress intensity factor at fracture is called the critical stress intensity factor, $K_C$.

Key points regarding $K_C$ are summarized as follows,

1. It is an experimental value derived from the theoretical $K$ equation.
2. As shown in Fig. 6.1a, $K_C$ varies with specimen thickness because the stress state changes with thickness. In other words, a plane stress state occurs in a thin specimen, while a plane strain state occurs in a thick specimen.
3. As shown in Fig. 6.1a, $K_C$ reaches a constant minimum value, $K_{IC}$ in the plane strain state.

The minimum value of $K_C$ in Mode I is called plane strain fracture toughness, $K_{IC}$.

Understanding the terminology of $K_{IC}$ is useful for the following reasons,

1. Since it is obtained from the specimen under the plane strain state, it has 'plane strain' on the terminology.
2. As it is determined at the point of fracture, the term includes 'fracture.'
3. Because it represents the material's resistance to fracture, it is termed 'fracture toughness.'

$K_{IC}$ is a material constant that must be obtained using a standard method. A high $K_{IC}$ value indicates that fracture occurs only when the stress near the crack tip is high, meaning the material is less prone to fracture.

 
K. Y. Lee, *Introduction to Elasticity, Fracture and Fatigue*,
https://doi.org/10.1007/978-981-96-3600-6_6

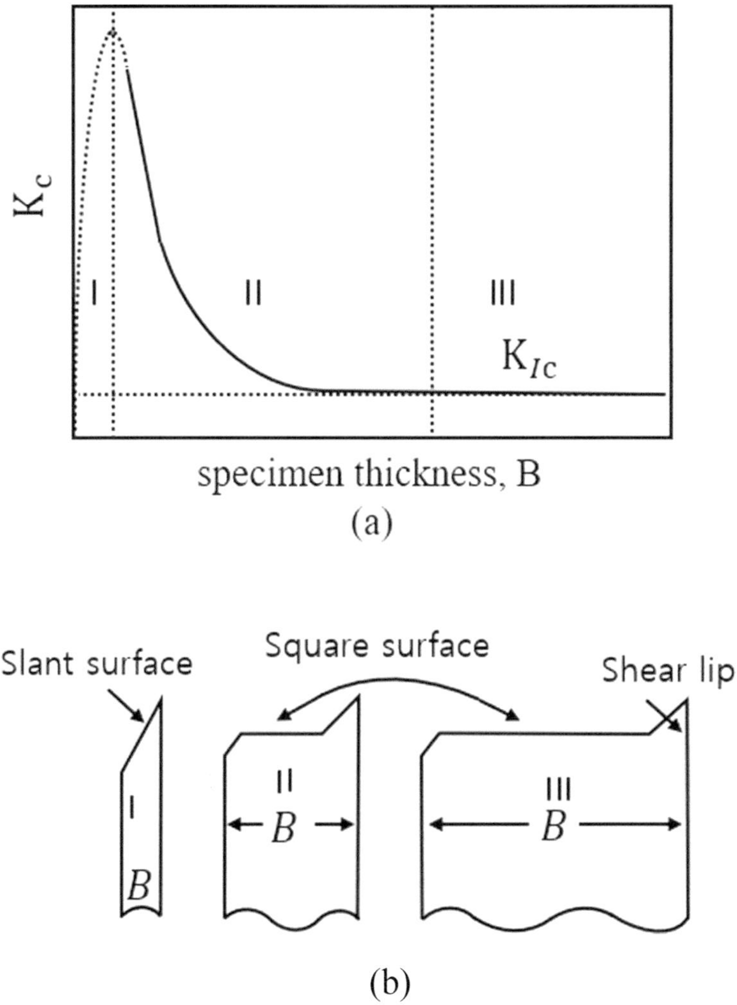


**Fig. 6.1** Variation of critical stress intensity factor

Figure 6.1b illustrates how the fracture surface shape varies with specimen thickness. In a thin specimen under plane stress, the fracture surface is slanted, while in a thick specimen under plane strain, it has a square shape. Notably, even in thick specimens, the sides of the fracture surface, which are in a plane stress state, show a slanted shear lip.

It is crucial to understand the difference between Figs. 6.2 and 6.3 to grasp the distinction between conventional design and fracture mechanics design.

Fig. 6.2 Specimen of different thickness without a crack

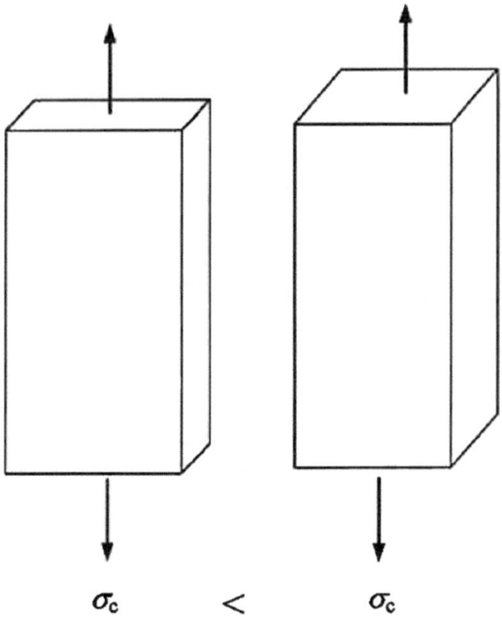


Fig. 6.3 Specimen of different thickness with a crack

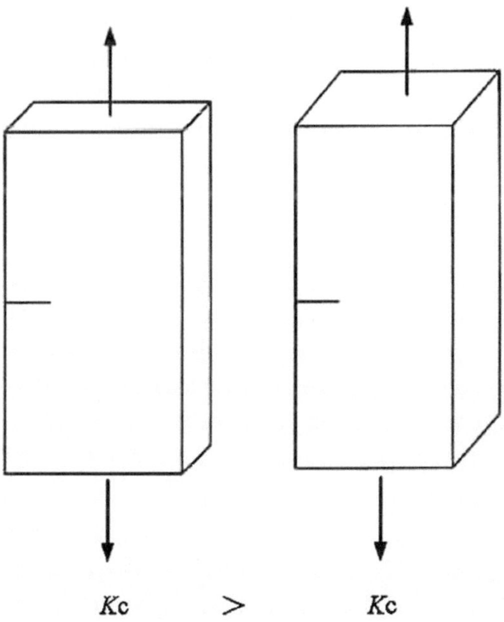

In conventional design, if a structure is crack-free throughout its service life, a thicker structure is preferable for safety, as shown in Fig. 6.2. However, micro-cracks (which are not classified as cracks in this context) are always present in materials. These micro-cracks grow over time and eventually develop into macro-cracks, which are considered cracks in this book. Therefore, the message of Fig. 6.3 is significant: it warns that a thick structure with a crack is more susceptible to fracture than a thin structure with a crack.

***Example 6.1*** Explain the reason for the transition in fracture surface appearance as shown in Fig. 6.1b.

**Solution**

When tensile force is applied longitudinally to a specimen in a tension test, the side surfaces of the specimen are always stress-free. Therefore, if the specimen is thin enough, it is completely in a plane stress state. However, if the specimen is thick enough, the mid-section exhibits a plane strain state, while the side surfaces remain in a plane stress state. Based on the concept that the plastic zone size in the plane stress state is larger than that in the plane strain state, we can draw Fig. 6.4. From Fig. 6.4, we can observe that a thin specimen does not exhibit a plane strain state in the mid-section. Consequently, the fracture surface of a thin specimen displays a slanted surface, as shown in Fig. 6.1b. This slanted surface resembles a mirror and is very sharp, referred to as a shear lip. However, even if a thick specimen has a plane strain state in the mid-section, the side surfaces remain in a plane stress state. Therefore, the fracture surface at the side surfaces forms a shear lip, while the mid-section shows a square fracture surface, as depicted in Fig. 6.1b for a thick specimen.

***Example 6.2*** Why do the fracture surfaces in the region of the plane stress state show a slant surface, while those in the region of the plane strain state show a rough and square surface?

**Solution**

The slant fracture surface of the shear lip forms at the boundary between the plastic zone and the elastic zone, as shown in Figs. 6.4 and 6.5, resulting in a slanted shear lip. In a thick specimen, the mid-section is in the plane strain state. Because the plastic zone in the plane strain state is small, the crack front propagates along the top line of the plastic zone at one moment and along the bottom line at another. Consequently, the fracture surface in the plane strain state becomes rough.

## 6.2 Plane Strain Fracture Toughness Testing

The plane strain fracture toughness, $K_{IC}$ is the minimum value of $K_C$. $K_{IC}$ is a material constant and should be obtained by a standard method.

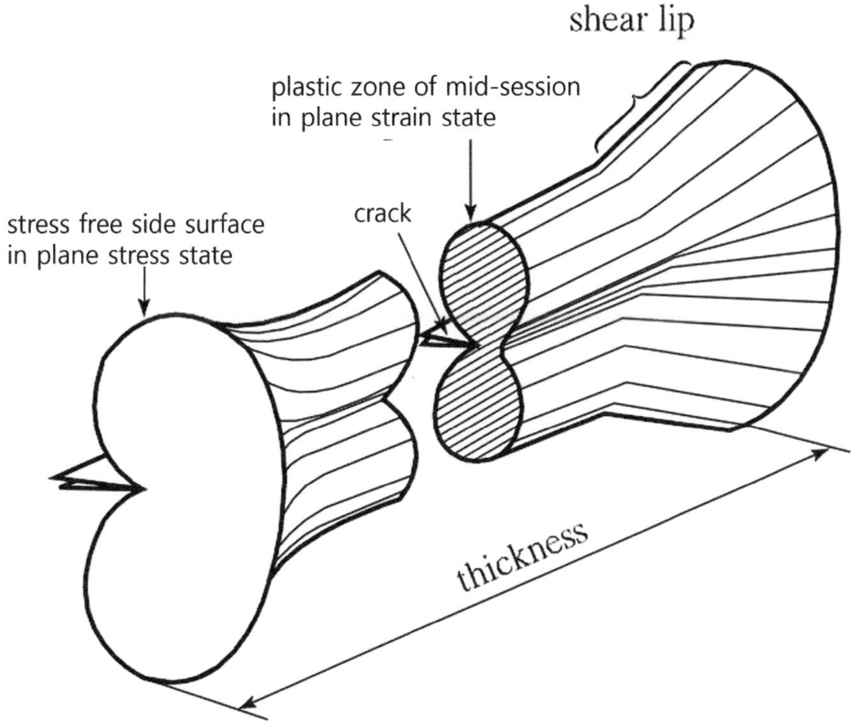


**Fig. 6.4** Three-dimensional plastic zone

**Fig. 6.5** Fracture surface

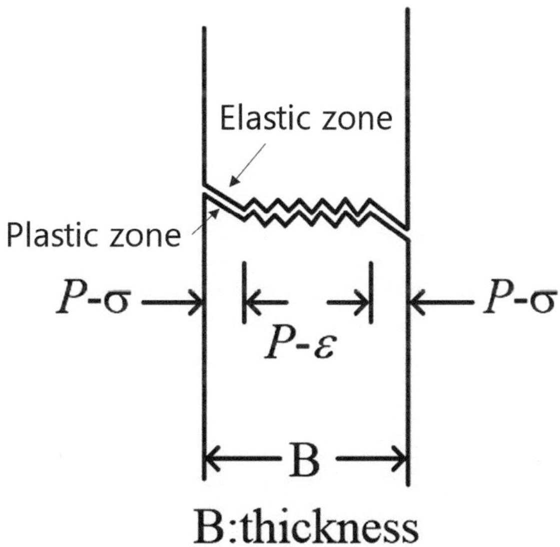

***Example 6.3*** Why should $K_{IC}$ be used rather than $K_C$ in design?

**Solution**

$K_{IC}$ is the minimum of $K_C$.

$$\therefore K_C > K_{IC}$$

According to Eq. 5.10,

$$Y\left(\frac{a}{W}\right)\sigma_C\sqrt{a} > Y\left(\frac{a}{W}\right)\sigma_{IC}\sqrt{a} \;\therefore\; \sigma_C > \sigma_{IC}$$

It means that $\sigma_{IC}$ is the minimum value of $\sigma_C$.

Therefore, using $K_{IC}$ instead of $K_C$ implies using the minimum allowable stress $\sigma_{IC}$ rather than $\sigma_C$. In other words, it is safer against fracture due to cracking when using $K_{IC}$ rather than $K_C$. This is the reason for measuring $K_{IC}$.

Since plane strain fracture toughness is a material constant, it should be measured by a standard method. In this book, ASTM E399-83, *Standard Test Method for Plane-Strain Fracture Toughness of Metallic Materials*, is adopted. '83' refers to the revision year. Whenever you test a material constant, you must read and adhere to the corresponding standard. In other words, you must not base any determination solely on your judgment without consulting the standard. If the standard does not provide answers to your questions, then you may proceed with your own approach. The steps for measuring $K_{IC}$ are as follows,

Step 1. Select one of several standard specimens shown in Fig. 6.6.

$B$ is the thickness of the specimen, and $a$ is the crack length. Depending on $B$ and $a$ (usually $a = B$), other dimensions are automatically determined.

Step 2. Determine the specimen size according to Table 6.1.

In Table 6.1, $\sigma_{ys}$ is yield stress and $E$ is Young's modulus.

Step 3. Carefully create a notch. When making a notch, ensure sufficient cooling to prevent changes in material properties due to heat. Figure 6.7 shows the shape of a notch.

Step 4. Generate a pre-fatigue crack from the notch using a Universal Testing Machine (UTM). The magnitude of the fatigue load and the length of the pre-fatigue crack should be determined according to the standard. Exercise patience when creating the pre-fatigue crack, as initiating it from the notch tip can be challenging. Applying a large fatigue load to speed up crack initiation risks breaking the specimen or creating an excessively large plastic zone ahead of the pre-fatigue crack.

Step 5. Install a COD (Crack Opening Displacement) gauge at the notch mouth, as shown in Fig. 6.8.

The COD gauge measures the opening displacement of the notch mouth under applied load.

Step 6. Load the specimen.

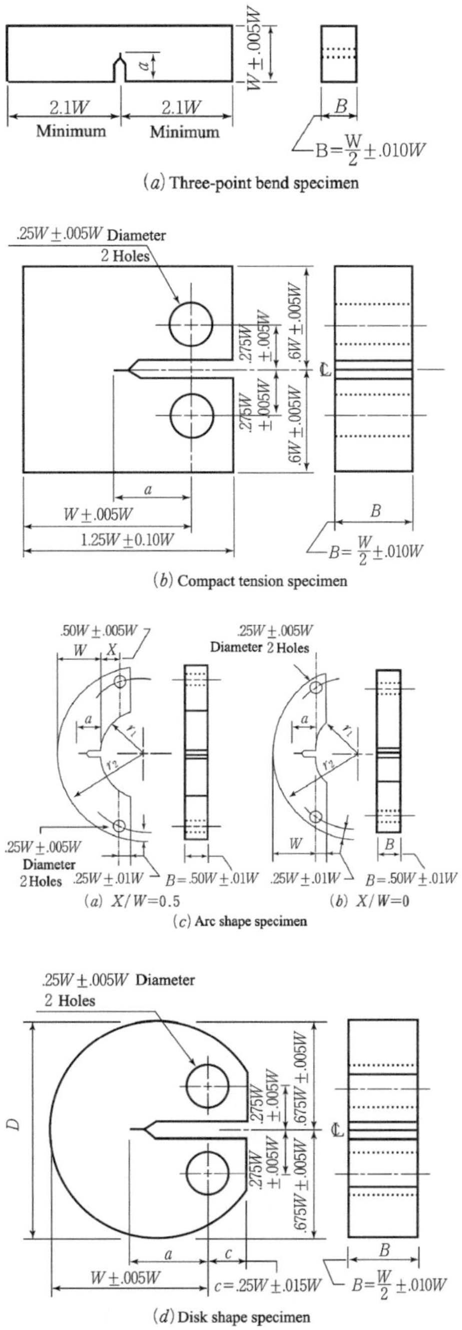


**Fig. 6.6** Standard specimens for testing plane strain fracture toughness (Reprinted with permission from ASTM E399-83, *Standard Test Method for Plane-Strain Fracture Toughness of Metallic Materials,* copyright ASTM International, 100 Barr Harbor Drive, West Conshohocken, PA 19428)

**Table 6.1** Determination of specimen thickness for plane strain fracture toughness testing (Reprinted with permission from ASTM E399-83, *Standard Test Method for Plane-Strain Fracture Toughness of Metallic Materials,* copyright ASTM International, 100 Barr Harbor Drive, West Conshohocken, PA 19428)

| Recommend minimum thickness (B) and crack length (a) | | |
|---|---|---|
| $\sigma_{ys}/E$ | in. | mm |
| 0.0050–0.0057 | 3 | 75 |
| 0.0057–0.0062 | 2 1/2 | 63 |
| 0.0062–0.0065 | 2 | 50 |
| 0.0065–0.0068 | 1 3/4 | 44 |
| 0.0068–0.0071 | 1 1/2 | 38 |
| 0.0071–0.0075 | 1 1/4 | 32 |
| 0.0075–0.0080 | 1 | 25 |
| 0.0080–0.0085 | 3/4 | 20 |
| 0.0085–0.0100 | 1/2 | 12 1/2 |
| 0.0100 above | 1/4 | 6 1/2 |

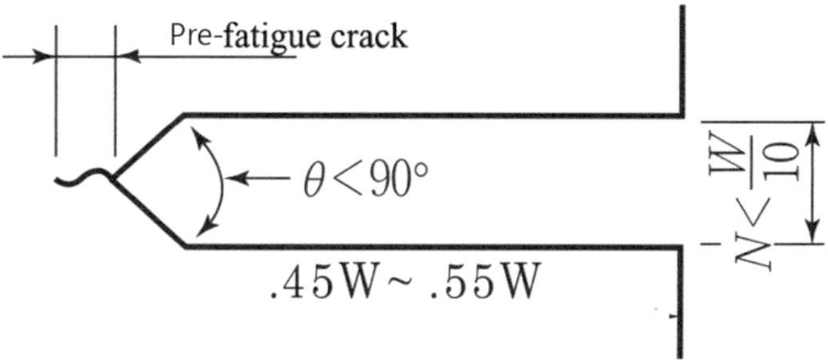


**Fig. 6.7** Notch for plane strain fracture toughness testing (Reprinted with permission from ASTM E399-83, *Standard Test Method for Plane-Strain Fracture Toughness of Metallic Materials,* copyright ASTM International, 100 Barr Harbor Drive, West Conshohocken, PA 19428)

Under the applied load, *X–Y* recorder attached to UTM generates a graph as shown in Fig. 6.9. Depending on the specimen material, one of the three types of load versus COD curves illustrated in Fig. 6.9 can be obtained.
Step 7. Draw a straight line with 5% down slope from the original slope shown in Fig. 6.9 and label the intersection point of this line with the load versus COD curve as $P_5$. Identify the maximum load between the lines with the original initial slope and the 5% down slope, and label it as $P_Q$ as shown in Fig. 6.9.
Step 8. Calculate $K_Q$ by substituting $P_Q$ into the $K$ equation corresponding to the specimen.

1. For the three-point bend specimen

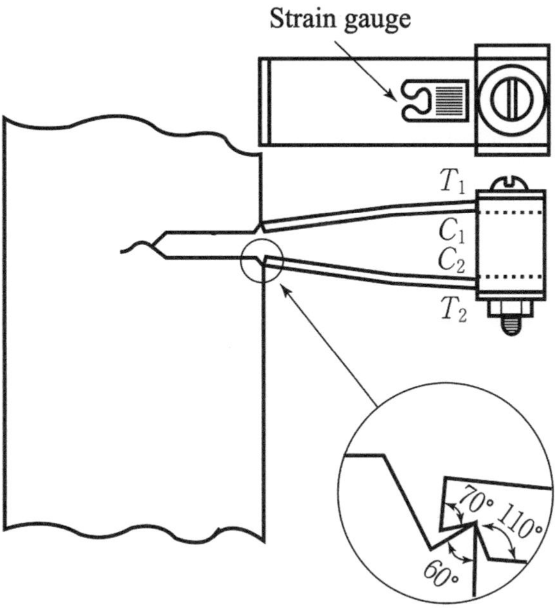


**Fig. 6.8** Installation of COD gauge (Reprinted with permission from ASTM E399-83, *Standard Test Method for Plane-Strain Fracture Toughness of Metallic Materials,* copyright ASTM International, 100 Barr Harbor Drive, West Conshohocken, PA 19428)

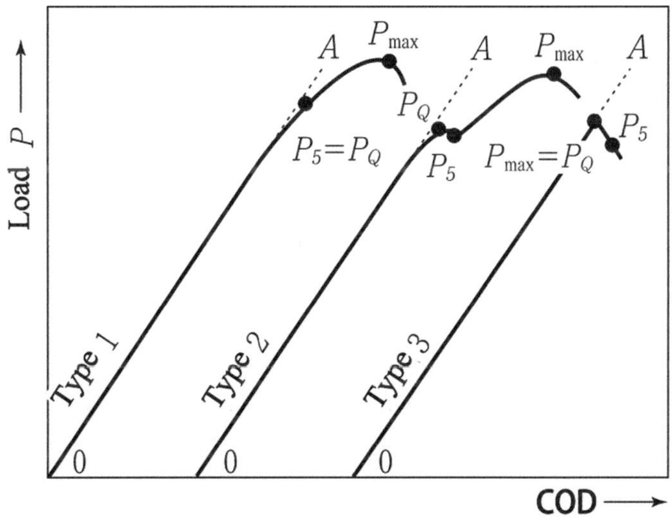


**Fig. 6.9** Load versus COD (Reprinted with permission from ASTM E399-83, *Standard Test Method for Plane-Strain Fracture Toughness of Metallic Materials,* copyright ASTM International, 100 Barr Harbor Drive, West Conshohocken, PA 19428)

$$K_Q = (P_Q S/BW^{3/2}) f(a/W)$$

$$f(a/W) = \frac{3(a/W)^{1/2}\left[1.99 - (a/W)(1 - a/W)\left(2.15 - 3.93a/W + 2.7a^2/W^2\right)\right]}{2(1 + 2a/w)(1 - a/W)^{3/2}} \tag{6.1}$$

where $S$ is span length.

2. For the compact tension specimen

$$K_Q = \left(P_Q/BW^{1/2}\right) f(a/W)$$

$$f(a/W) = \frac{(2 + a/W)(0.886 + 4.64a/W - 13.32a^2/W^2 + 14.72a^3/W^3 - 5.6a^4/W^4)}{(1 - a/W)^{3/2}} \tag{6.2}$$

3. For the arc shape specimen

$$\begin{aligned} K_Q &= (p_Q/BW^{1/2})[3X/W + 1.9 + 1.1a/W] \\ &\quad \times [1 + 0.25(1 - a/W)^2(1 - r_1/r_2)] f(a/W) \\ f(a/W) &= [(a/W)^{1/2}/(1 - a/W)^{3/2}] \\ &\quad \times [3.74 - 6.30a/W + 6.32a^2/W^2 - 2.43a^3/W^3]. \end{aligned} \tag{6.3}$$

4. For the disk shape specimen

$$K_Q = (P_Q/BW^{1/2}) f(a/W)$$

$$f(a/W) = \frac{(2 + a/W)(0.76 + 4.8a/W - 11.58a^2/W^2 + 11.43a^3/W^3 - 4.08a^4/W^4)}{(1 - a/W)^{3/2}} \tag{6.4}$$

Step 9. Introduce $K_Q$ into Eq. 6.5

$$B, a \geq 2.5(K_Q/\sigma_{ys})^2 \tag{6.5}$$

If this equation is satisfied, designate $K_Q$ as $K_{IC}$. Otherwise, repeat the test with a thicker specimen. Equation 6.5 is used to verify whether the plastic zone near the crack tip is sufficiently small for $K$ to be applicable.

***Example 6.4*** Why should a fatigue load larger than the standard load not be applied when creating a pre-fatigue crack?

**Solution**

$K_{IC}$ is a parameter in the linear elastic fracture mechanics (LEFM). LEFM does not permit a plastic zone at the crack tip, as elasticity theory considers only elastic deformation. However, to apply the stress intensity factor $K$ to real materials that

**Table 6.2** The plane strain fracture toughness $K_{IC}$ and yield stress $\sigma_{ys}$

| Materials | $K_{IC}\left(\text{ksi}\sqrt{\text{in}}\right)$ | $\sigma_{ys}$(ksi) |
|---|---|---|
| 7178-T651 | 21 | 83 |
| 7178-T7651 | 30 | 71 |
| Ti-6Al-4 V | 105 | 132 |
| Ti-6Al-4 V | 50 | 150 |
| 4340 | 90 | 125 |
| 4340 | 55 | 220 |
| H-11 | 35 | 260 |
| H-11 | 25 | 300 |
| 350 Maraging | 50 | 225 |
| 350 Maraging | 35 | 325 |

yield under high stress, a small plastic zone ahead of the crack tip is permissible. If a high fatigue load is applied when creating a pre-fatigue crack, an excessive plastic zone may form ahead of the crack tip, which is not allowable.

***Example 6.5*** What is the physical meaning of Eq. 6.5?

**Solution**

Recalling Eq. 5.25, we understand that Eq. 6.5 imposes a restriction on the plastic zone size ahead of the crack.

Table 6.2 presents data on plane strain fracture toughness $K_{IC}$ and yield stress $\sigma_{ys}$ for various materials.

Figure 6.10 shows the relationship between plane strain fracture toughness $K_{IC}$ and yield stress $\sigma_{ys}$. Notably, there is an inverse relationship between $K_{IC}$ and yield stress $\sigma_{ys}$. This has significant implications: a high-strength material with a high yield stress has a low plane strain fracture toughness. This means that when a high-strength material is chosen in design, it may be more prone to fracture due to its low plane strain fracture toughness.

## 6.3 Fracture Mechanics Design

The first step in fracture mechanics design depends on identifying the most critical parameter among crack length, applied stress, and material.

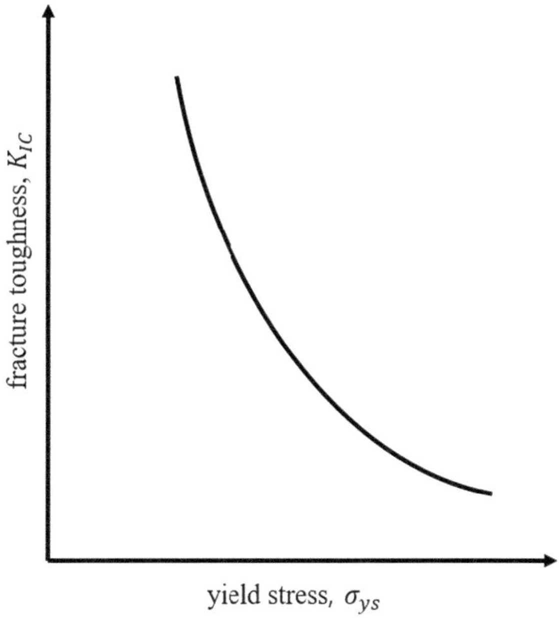


**Fig. 6.10** Relationship between plane strain fracture toughness $K_{IC}$ and yield stress

### *6.3.1 In the Case of Reactor Design*

1. First, select a material, as resistance to radiation-induced corrosion is essential.
2. Measure $K_{IC}$ for the selected material.
3. Choose an allowable crack length that limits radiation leakage and ensures detectability via Nondestructive Testing (NDT).

In this scenario, the crack length should be very small compared to the body size. Therefore, we use the $K$ equation for a center crack in an infinite body.

$$\therefore K = \sigma\sqrt{\pi a}$$

At fracture, we can express $K_{IC} = \sigma_{IC}\sqrt{\pi a}$. For the safest design, we use $K_{IC}$ instead of $K_C$. Introduce the measured $K_{IC}$ and the selected allowable crack length $a$ to obtain $\sigma_{IC}$. The pressure in the reactor is fixed, as it operates effectively only at a specific pressure.

$\therefore$ The cross-sectional area is determined based on the definition of stress.

$\therefore$ The thickness and diameter of the reactor can then be calculated from the determined cross-sectional area.

### 6.3.2 In Case of the Design of the Airplane

1. First, select a lightweight material to reduce the airplane's overall weight.
2. Measure $K_{IC}$ for the chosen material
3. Determine an allowable fracture stress, as high stress can occur during takeoff and landing.

Since the allowable crack length is very small, comparing to the cracked body, we can assume a center crack in the infinite body.

Therefore,

$$\therefore K = \sigma\sqrt{\pi a}$$

At fracture, we can express $K_{IC} = \sigma_{IC}\sqrt{\pi a}$. For the safest design, $K_{IC}$ is used instead of $K_C$. Introduce the measured $K_{IC}$ and the selected allowable fracture stress $\sigma_{IC}$ to calculate $a_c$.

4. The value $a_c$ should be detectable by NDT and larger than the radius of rivet head as a crack hidden under the rivet head may go undetected by NDT. If $a_c$ is too small to be detectable by NDT, select a material with a higher $K_{IC}$.

***Example 6.6*** Consider a plate without a crack under an applied force. Determine the plate thickness required to prevent yielding and fracture, assuming the plate width is given.

**Solution**

To prevent yielding and fracture, both the conventional design method using yield stress and the fracture mechanics design method using plane strain fracture toughness are considered.

**Conventional design method using yield stress $\sigma_{ys}$.**

The maximum stress $\sigma_{max}$ in the plate should be less than yield stress $\sigma_{ys}$ to avoid yielding

$$\sigma_{max} < \sigma_{ys}$$

$$\therefore \quad \frac{P_{max}}{A} < \sigma_{ys}$$

where $P_{max}$ is the applied force and $A$ is the cross-sectional area.

Since $A = BW$ where $B$ and $W$ are the thickness and width of the plate, respectively,

$$\frac{P_{max}}{BW} < \sigma_{ys}$$

From this equation, the thickness $B$ can be determined.

**Fracture mechanics design using $K_{IC}$ to avoid fracture.**

It is assumed that a center crack may develop in the plate over time, even if none is currently present. Due to this assumption, fracture mechanics design is considered highly safe against fracture. Since the allowable crack length is typically very small compared to the width of the plate, we assume a center crack in an infinite body.

$$\therefore \quad K = \sigma\sqrt{\pi a}$$

Introduce the allowable crack length $a_{all}$ and $K_{IC}$.

$$K_{IC} = \sigma_{IC}\sqrt{\pi a_{all}}$$

Obtain $\sigma_{IC}$.

$\sigma_{IC}$ must be higher than $\sigma_{max}$.

Therefore,

$$\sigma_{max} < \sigma_{IC}$$
$$\frac{P_{max}}{BW} < \sigma_{IC}$$

Obtain the value of the thickness $B$.

Finally, to prevent yielding and fracture during use, the plate thickness should be set to a value greater than those obtained from both the conventional design and the fracture mechanics design.

# Chapter 7
# Fatigue Fracture

**Abstract** This chapter briefly introduces the standard testing method for fatigue crack growth rate. The difference between Paris' law and Forman's law is explained. How to search why a crack occurred in the structure and how to make fatigue crack growth retard also introduced.

Note that fatigue fracture and fatigue failure are different. Fatigue fracture refers to destruction due to cracks under fatigue load, while fatigue failure refers to destruction without cracks under fatigue load. In this book, any reference to a crack denotes a macro-crack, not the micro-cracks that are always initially present in materials.

## 7.1 Fatigue Crack Growth Rate

Figure 7.1 shows crack growth with increasing cycles when a body with an initial crack length $a_0$ is subjected to fatigue loading with a mean stress $\sigma_m = 0$.

Note that crack growth rate $\frac{\mathrm{d}a}{\mathrm{d}N}$ depends on the stress range $\Delta\sigma$ and crack length $a$.

In other word,

$$\mathrm{d}a/\mathrm{d}N = f(\Delta\sigma, a) \tag{7.1}$$

Since the stress intensity factor is a function of stress and crack length, Eq. 7.1 can be written as follows,

$$\mathrm{d}a/\mathrm{d}N = f(\Delta K) \tag{7.2}$$

where $\Delta K$ is called the range of stress intensity factor.

$\Delta K$ is defined as follows,

$$\Delta K = K_{\max} - K_{\min} \tag{7.3}$$



K. Y. Lee, *Introduction to Elasticity, Fracture and Fatigue*,
https://doi.org/10.1007/978-981-96-3600-6_7

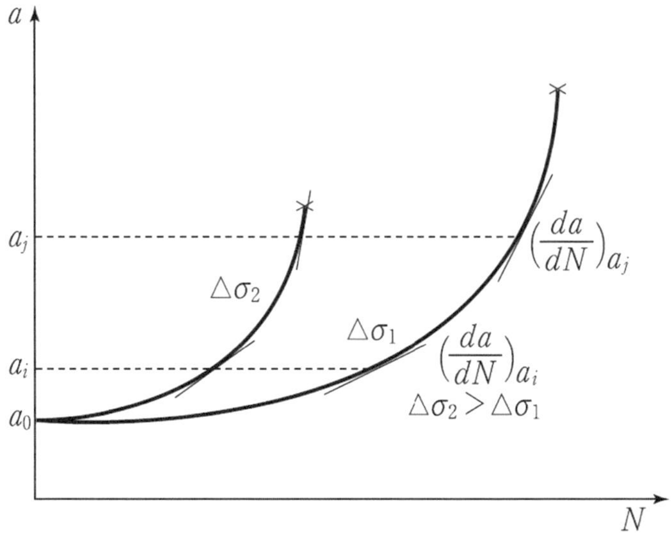


**Fig. 7.1** Fatigue crack growth

The experimental result of Eq. 7.2 is shown in Fig. 7.2.

$\Delta K_{\rm th}$ in Fig. 7.2 is called the range of threshold stress intensity factor. There is no fatigue crack growth in $\Delta K$ less than $\Delta K_{\rm th}$.

***Example 7.1*** How can Fig. 7.2 be obtained using a single specimen?

**Fig. 7.2** Dependence between crack growth rate and the range of stress intensity factor

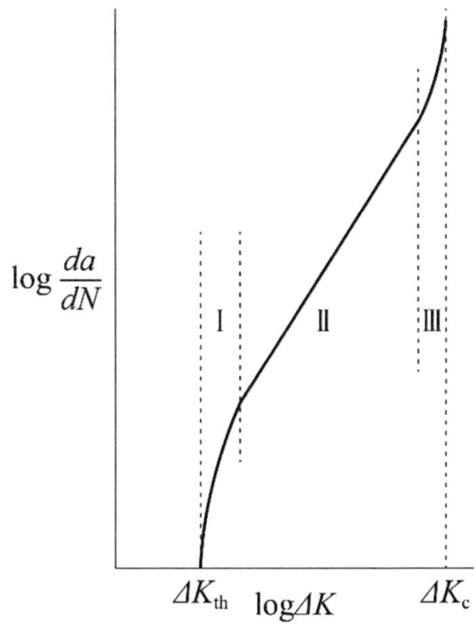

**Solution**

Obtaining Fig. 7.2 with a single specimen, starting from an initial crack length $a_0$ requires several steps.

Step 1. Determine $\Delta\sigma$ that initiates crack growth.
This step requires significant time, as $\Delta\sigma$ must be gradually increased from a low value. Once $\Delta\sigma$ reaches $\Delta\sigma_{\mathrm{th}}$ initiating crack growth, calculate $\Delta K_{\mathrm{th}}$ using the $K$ equation for the specimen's shape.
Step 2. Apply $\Delta\sigma_{\mathrm{th}}$ continuously.
Step 3. Measure the crack length and cycle number at regular time interval.
Step 4. Plot the crack length against cycle number measurements.
Step 5. At the several crack lengths, determine $\mathrm{d}a/\mathrm{d}N$ from the plot and calculate $\Delta K$.
Step 6. Plot $\mathrm{d}a/\mathrm{d}N$ versus $\Delta K$ to obtain Fig. 7.2.

Figure 7.2 displays three zones. Zones I and III exhibit non-linear behavior, while Zone II shows linear behavior.

## 7.2 Paris' Law

The relationship between $\mathrm{d}a/\mathrm{d}N$ and $\Delta K$ in Zone II in Fig. 7.2 can be expressed as follows,

$$\mathrm{d}a/\mathrm{d}N = A(\Delta K)^m \tag{7.4}$$

where $A$ and $m$ are experimental constants that depend on $R$.

We refer to Eq. 7.4 as Paris' Law.

To calculate fatigue life using Paris' Law, Eq. 7.4 is modified as follows,

$$\Delta N = \int_{a_0}^{a_f} \frac{1}{A(\Delta K)^m} \mathrm{d}a \tag{7.5}$$

where $a_0$ is the initial crack length and $a_f$ is the fatigue crack length at a specific moment.

$\Delta K$ can be expressed by replacing $K$ with $\Delta K$ and $\sigma$ with $\Delta\sigma$ in the general form of stress intensity factor, as follows,

$$\Delta K = Y\left(\frac{a}{W}\right)\Delta\sigma\sqrt{a} \tag{7.6}$$

By substituting Eq. 7.6 into Eq. 7.5,

$$\therefore \Delta N = \frac{1}{A(\Delta\sigma)^m} \int_{a_0}^{a_f} \frac{1}{(Y\sqrt{a})^m} \mathrm{d}a \tag{7.7}$$

This equation provides the number of fatigue cycles required for crack growth from an initial crack length $a_0$ to $a_f$. Although Paris' Law applies strictly to Zone II, as shown in Fig. 7.2, it is noted that Eq. 7.7 can be used as an approximation for fatigue life up to Zone III.

***Example 7.2*** Initially, there was no crack in the plate. However, assume that an edge crack was later found perpendicular to the applied fatigue load. Determine the reason for the appearance of this crack.

**Solution**

There are several possible reasons for the appearance of the crack:

1. An original defect at the cracked location, such as a sharp notch or inclusion.
2. Corrosion due to the presence of a corrodible chemical.

But in this example, let's determine whether the operator maintained the allowable stress range during operation. To do this, calculate the applied $\Delta\sigma$ value. Since the plate has a crack, use the $K$ equation corresponding to the cracked body under the applied load. Calculating $\Delta\sigma$ for the plate involves several steps, as follows,

Step 1. Separate the cracked body into two parts to observe the crack surface.
Step 2. Examine the fracture surface.
If you observe continuous striations through the thickness of the fracture surface, the crack can be attributed to fatigue loading. If the crack is due to fatigue load, you will need the measurement results for Paris' Law.
Step 3. Measure the spacing between the striations at several locations along the direction of crack growth.
The gap between two striations represents crack growth per one cycle, $da/dN$. Using the Paris' law measurements, obtain $\Delta K$ from $da/dN$ at various crack locations.
Step 4. Using the $\Delta K$ equation, calculate $\Delta\sigma$ at the crack locations and compare these values with the allowable stress range. If $\Delta\sigma$ is less than the allowable stress range, we can conclude that the operator is not responsible for the crack's occurrence.

As illustrated in this example, the fracture surface is crucial for identifying the cause of cracking. Therefore, the fracture surface should be preserved carefully to facilitate this analysis. A specialist who investigates the cause of cracking is called a fracture analyst.

## 7.3 Forman's Law

Forman's law is expressed as follows,

$$\frac{da}{dN} = \frac{C(\Delta K)^n}{(1 - R)K_c - \Delta K} \tag{7.8}$$

where $C$ and $n$ are experimental constants.

However, note that $C$ and $n$ do not depend on $R$, according to experimental results.

***Example 7.3*** Why can Forman's Law be applied up to the point of crack propagation?

**Solution**

In Forman's law, the load ratio $R$ is defined as follows,

$$R = \frac{\sigma_{min}}{\sigma_{max}} \tag{7.9}$$

Therefore, by multiplying both the numerator and denominator in Eq. 7.9 by the same term, it can be rewritten as follows,

$$R = \frac{Y\left(\frac{a}{W}\right)\sigma_{\min}\sqrt{a}}{Y\left(\frac{a}{W}\right)\sigma_{\max}\sqrt{a}} = \frac{K_{\min}}{K_{\max}}$$

Therefore,

$$R = \frac{K_{min}}{K_{max}} \tag{7.10}$$

By substituting Eq. 7.10 into the denominator of Forman's Law, it can be expressed as follows,

$$(1 - R)K_c - \Delta K = \left(\frac{K_{\max} - K_{\min}}{K_{\max}}\right)K_c - \Delta K = \frac{\Delta K}{K_{\max}}K_c - \Delta K$$

Since the crack propagates at $K_{max}$, $K_{max}$ must be $K_c$ at the propagation moment. Therefore, at the propagation moment,

$$\frac{\Delta K}{K_{\max}}K_c - \Delta K = 0$$

This results in an infinite crack growth rate in Eq. 7.8. Therefore, we can conclude that Forman's Law is applicable up to the moment of propagation.

## 7.4 Fatigue Crack Growth Rate Testing

To measure the fatigue crack growth rate, we use ASTM E647-88a, *Standard Test Method for Measurement of Fatigue Crack Growth Rates*.

The test sequence is outlined as follows,

Step 1. Select the specimen shape from Fig. 7.3.
Step 2. Create a notch.
Step 3. Generate a pre-fatigue crack.
Step 4. Begin the fatigue crack growth test by applying a fatigue load.
Step 5. Measure a fatigue crack length $a$ and fatigue cycle $N$. It is necessary to occasionally stop a UTM to measure the fatigue crack length $a$ and fatigue cycle $N$.

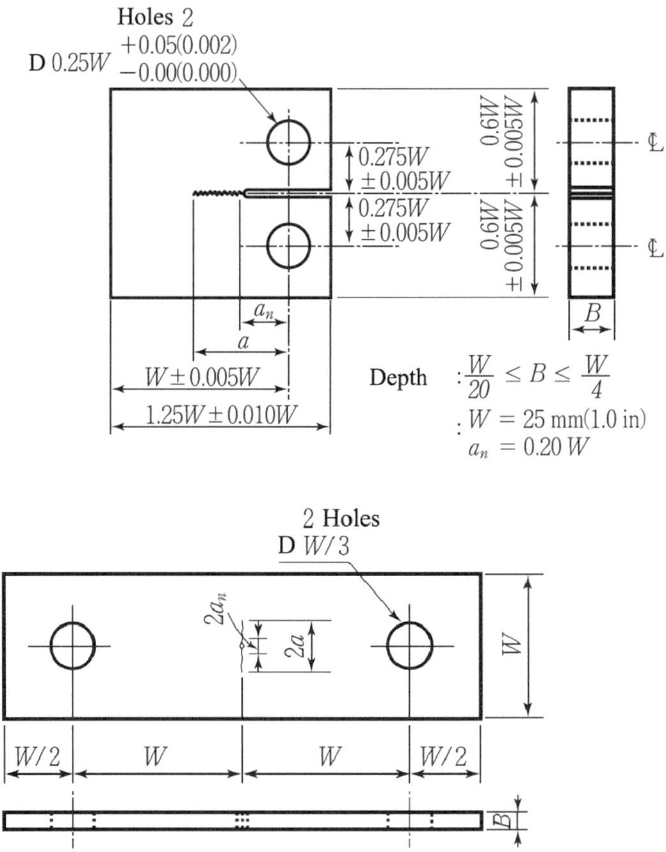


**Fig. 7.3** Specimen for fatigue crack growth rate testing (Reprinted with permission from ASTM E647-88a, *Standard Test Method for Measurement of Fatigue Crack Growth Rates*, copyright ASTM International, 100 Barr Harbor Drive, West Conshohocken, PA 19428)

Step 6. Using the measurements of fatigue crack length and fatigue cycle, calculate $\frac{da}{dN}$ using Eq. 7.11.

$$\left(\frac{da}{dN}\right)_{\overline{a}} = \frac{a_{i+1} - a_i}{N_{i+1} - N_i} \tag{7.11}$$

where $\overline{a} = \frac{1}{2}(a_{i+1} + a_i)$.
Step 7. Select the $\Delta K$ equation corresponding to the specimen and calculate $\Delta K$ value.

1. For a compact tension specimen

$$\begin{aligned}\Delta K = & \frac{\Delta P(2+\alpha)}{B\sqrt{W}} \frac{1}{(1-\alpha)^{3/2}} \\ & \times \left(0.886 + 4.64\alpha - 13.32\alpha^2 + 14.72\alpha^3 - 5.6\alpha^4\right) \\ & (\overline{a}/W \geq 0.2, \alpha = \overline{a}/W)\end{aligned} \tag{7.12}$$

2. For a center crack tension specimen

$$\Delta K = \frac{\Delta P}{B}\sqrt{\frac{\pi\alpha}{2W}\sec\frac{\pi\alpha}{2}} \tag{7.13}$$

$$(2\overline{a}/W < 0.95, \alpha = 2\overline{a}W)$$

Step 8. Plot the results of $da/dN$ and $\Delta K$.

## 7.5 Retardation of Fatigue Crack Growth

Figure 7.4 shows that the fatigue crack growth is retarded when a moderate peak fatigue load is applied during constant fatigue loading.

In Fig. 7.4, $N_p$ is the cycle number progressed under the effect of peak fatigue load, $N_d$ is the extension of fatigue life due to the peak fatigue load, and $a_p$ is the fatigue crack length grown under the effect of the peak fatigue load.

It is important to note that the peak fatigue load should be moderate. If it is too high, fracture propagation could occur at the moment when the peak fatigue load is applied.

***Example 7.4*** Why is the fatigue crack growth retarded by applying the peak load during constant fatigue loading?

**Solution**

Figure 7.5 shows the difference in plastic zone sizes under constant fatigue load without a peak load and with a peak fatigue load.

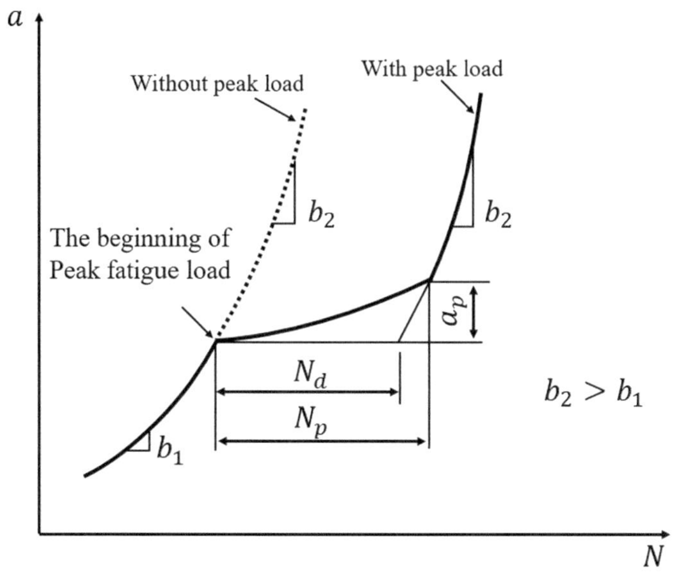


**Fig. 7.4** Fatigue Crack Growth Retardation

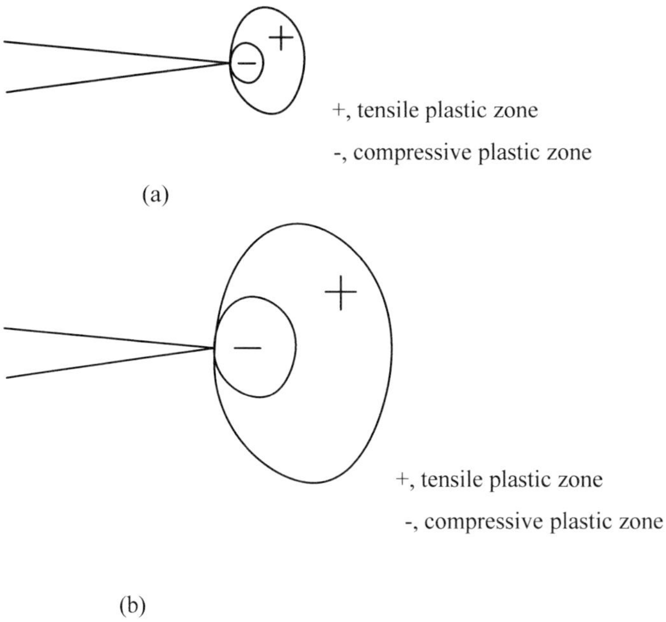


**Fig. 7.5** Plastic zone sizes under constant fatigue load without peak load (**a**) and with peak load (**b**)

The front of the crack tip under constant fatigue load, whether with or without a peak load, always has a tensile yield zone. The crack growth is not significantly affected by the size of the tensile yield zone, but is mainly influenced by the size of the compressive yield zone. Since the size of the compressive zone under constant fatigue load with a peak load is larger than under constant fatigue load without a peak load, the peak load causes retardation of fatigue crack growth.

It is important to note that the reason a compressive plastic zone occurs ahead of the crack tip, even though the minimum stress intensity factor is positive, is due to the impact effect caused by the rapid change from the maximum to minimum stress intensity factor.

# Part III
# Fatigue

# Chapter 8
# Fatigue

**Abstract** This chapter explains why fatigue data have scattering band, how to get constant hysteresis loop from random fatigue load, how to get fatigue life, how to get approximation S-N curves in metallic martials.

## 8.1 Fatigue Load

Figure 8.1 shows a constant stress fatigue load.

In Fig. 8.1, $\Delta\sigma$ is called stress range, $\sigma_a$ is stress amplitude (or alternating stress), $\sigma_m$ is mean stress and $R$ is load ratio. $\sigma_{\max}$ is maximum stress and $\sigma_{\min}$ is minimum stress in constant stress fatigue load.

These symbols are defined as shown in Fig. 8.1 as follows,

$$\begin{aligned} \Delta\sigma &\equiv \sigma_{\max} - \sigma_{\min} \\ \sigma_a &\equiv \frac{\sigma_{\max} - \sigma_{\min}}{2} \\ \sigma_m &\equiv \frac{\sigma_{\max} + \sigma_{\min}}{2} \\ R &= \frac{\sigma_{\min}}{\sigma_{\max}} \end{aligned} \tag{8.1}$$

Let's consider the case of the same $\sigma_a$ (or $\Delta\sigma$) but different $\sigma_m$ to understand the effect of $\sigma_m$ on the fatigue life in Fig. 8.2. When mean stress increases under the same stress amplitude, it is noted that fatigue life decreases.

It is noted that the graph of the S–N curve in Fig. 8.3a is a semi-log S–N curve for metallic materials, which consists of two straight lines. However, the S–N curve for nonmetallic materials shows a curve, as shown in Fig. 8.3b. In the S–N curve for metallic materials, the parallel line indicates the fatigue limit, which is typically defined as the stress amplitude at $10^6$ cycles. It is a material constant measured by standard method. Standard S–N curve is obtained under $\sigma_m = 0$. The fatigue limit has the physical meaning that a material will have an infinite life under stress amplitudes lower than this limit. However, since a S–N curve in nonmetallic materials shows a

 
K. Y. Lee, *Introduction to Elasticity, Fracture and Fatigue*,
https://doi.org/10.1007/978-981-96-3600-6_8

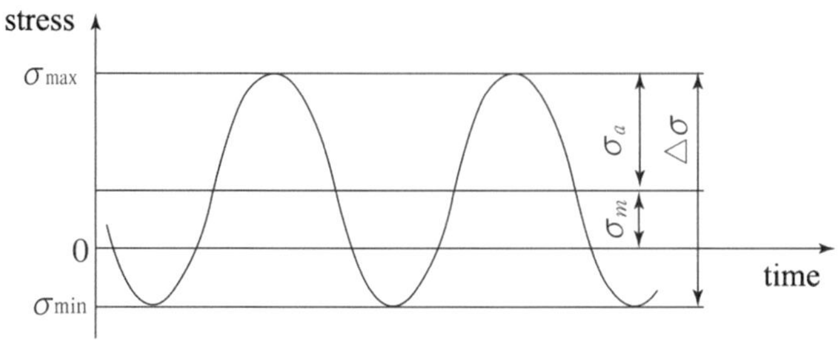


Fig. 8.1 Constant stress fatigue load

Fig. 8.2 Effect of mean stress

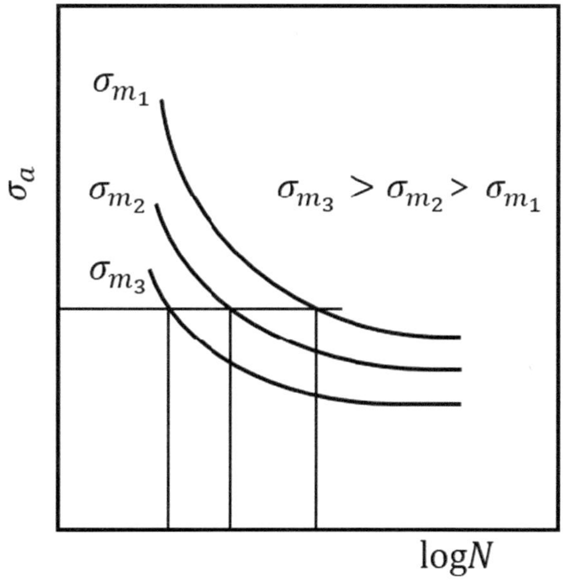


decreasing pattern as the cycle increases, it is predicted that they will eventually fail, even under low fatigue load levels. Therefore, nonmetallic materials do not actually have a fatigue limit. But for convenience in design, the stress amplitude at $10^6$ cycles is assumed to be the fatigue limit for nonmetallic materials.

In Fig. 8.3b, scattering bands are shown, whereas they are omitted in Fig. 8.3a. The scattering band indicates that different fatigue life data can be obtained from each specimen, even under the same stress amplitude.

The causes of the scattering of fatigue data are:

1. Differences in how the specimen is manufactured
2. Variations in surface roughness on the specimen
3. Differences in the microstructure of the specimen
4. Misalignment between the grip of the tester and the specimen.

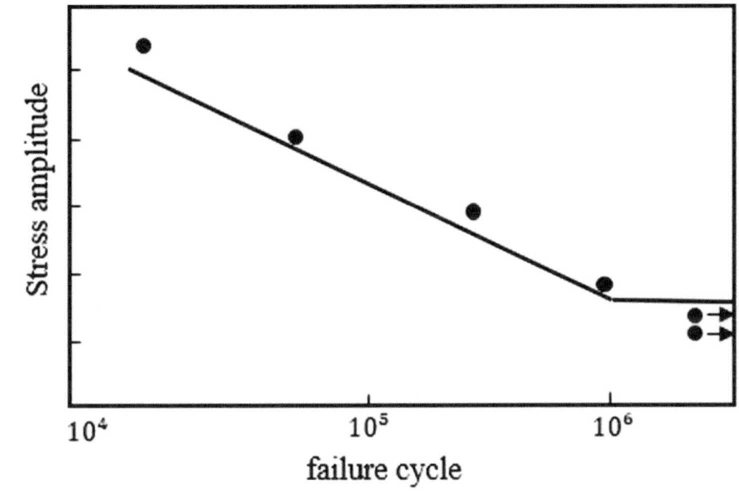


(a) S-N curve in metallic materials

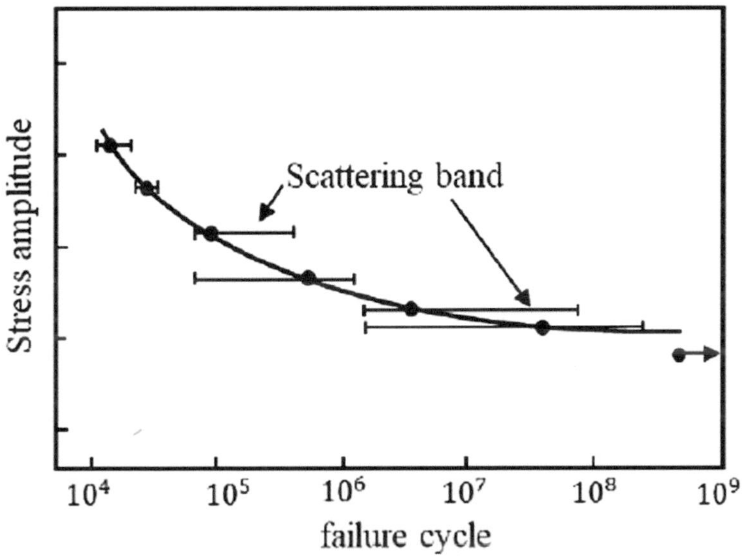


(b) S-N curve in nonmetallic materials

**Fig. 8.3** S–N curves, **a** S–N curve in metallic materials and **b** S–N curve in nonmetallic materials

To achieve proper alignment between the grip of the tester and the specimen, it is recommended to hang one end of a string from the ceiling and tie a weight to the other end. By observing from several different directions, you can check whether the string and specimen are parallel. The reason for the scattering band in the S–N curve is that, although the four causes mentioned above have minimal effect during a single cycle, their effects accumulate over the fatigue life cycles. Due to the differences in this accumulated effect in each specimen, the scatter band is produced. At lower stress amplitudes, the scattering band is wider because the accumulated effect is more pronounced.

***Example 8.1*** Can a monotonic tension test have a wide scattering band like a cyclic fatigue test?

**Solution**

No. Since there is no accumulated effect in a monotonic tension test, it is impossible to have such a wide scattering band as shown in a fatigue test.

When either a constant stress fatigue cycle or constant strain fatigue cycle is applied, as shown in Fig. 8.4b, c, a hysteresis loop, as shown in Fig. 8.4a, can be drawn. This kind of hysteresis loop is obtained in every cycle.

The area ABCDEF in Fig. 8.4 is called fatigue damage, which represents the plastic work done per unit volume. The larger this area, the greater the fatigue damage. Therefore, the fatigue life becomes shorter.

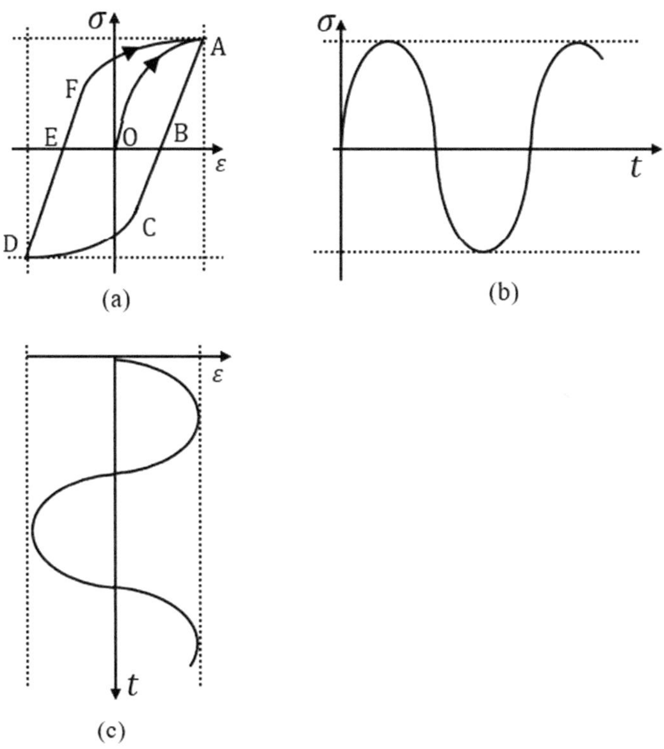


**Fig. 8.4** **a** Hysteresis loop obtained from either **b** constant stress fatigue cycle or **c** constant strain fatigue cycle

However, this type of hysteresis loop is continuously formed and changes slightly with each cycle. This means that, as the fatigue cycles continue, the material can either harden or soften. If the material hardens, we call it cyclic strain hardening, and if it softens, we call it cyclic strain softening.

***Example 8.2*** Let's say the fatigue load is given as shown in Fig. 8.5, and the material exhibits cyclic strain hardening. Draw the figure of strain versus time.

**Solution**

Since the material exhibits cyclic strain hardening, strain should decrease with the increase of cycle, as shown in Fig. 8.6.

***Example 8.3*** Let's say fatigue load is given like Fig. 8.7 and the material exhibits cyclic strain softening.

Draw the figure of stress versus time.

**Solution**

Since the material exhibits cyclic strain softening, stress should decrease as the cycles increase, as shown in Fig. 8.8.

Due to either cyclic strain hardening or cyclic strain softening under repeated fatigue load, the shape of the hysteresis loop should change slightly with each cycle.

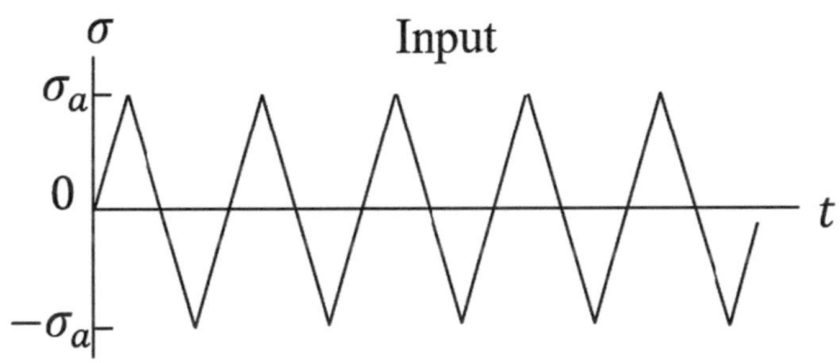


**Fig. 8.5** Constant stress cycle fatigue load as input

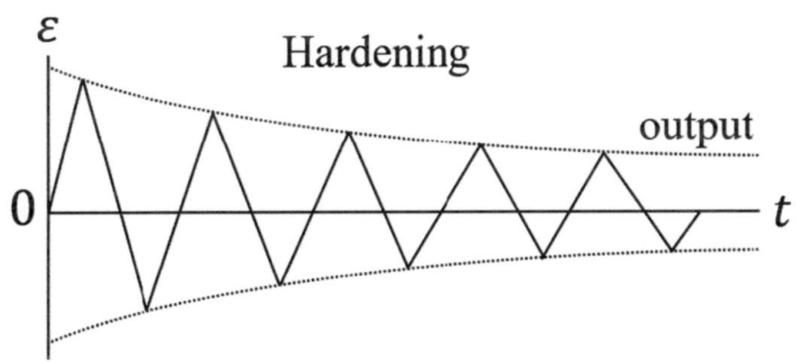


**Fig. 8.6** Cyclic strain hardening as output

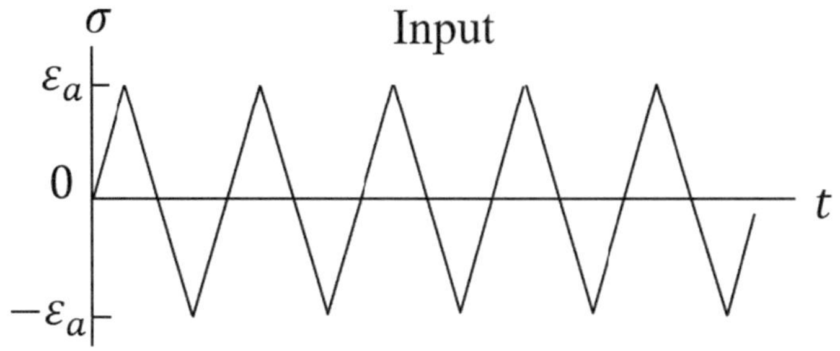


**Fig. 8.7** Constant strain cycle loads as input

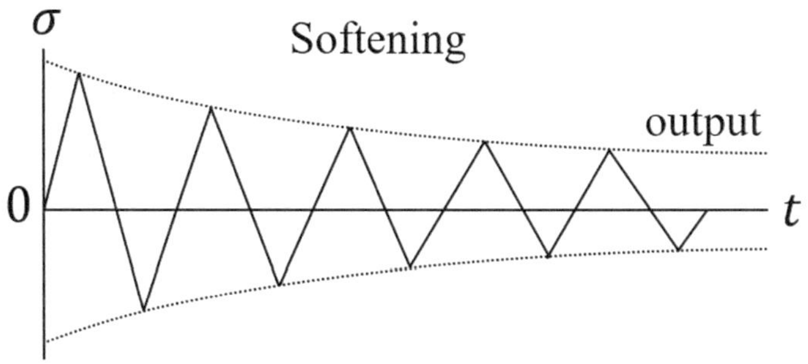


**Fig. 8.8** Cyclic strain softening as output

Figure 8.9 shows the gradual change in the hysteresis loop with each cycle. However, after many cycles, the loop no longer changes and becomes fixed. Such a fixed hysteresis loop is called a cyclically stabilized hysteresis loop. In Fig. 8.9, the thick solid curves represent cyclically stabilized hysteresis loops. It is noted that Fig. 8.9 is obtained under constant strain fatigue load. Figure 8.9a shows cyclic strain hardening, Fig. 8.9b shows cyclic strain hardening during the initial fatigue cycles, followed by cyclic strain softening afterward, and Fig. 8.9c shows cyclic strain softening.

## 8.2 Cyclic Stress and Strain Curve

We have already learned how to obtain a monotonic stress and strain curve in Mechanics of Materials or Solid Mechanics. Corresponding to that curve, we can obtain a cyclic stress and strain curve by applying a fatigue load. There are two methods to obtain this, one is the multi-specimen method, and the other is the single-specimen method. To obtain a cyclic stress and strain curve from multi-specimens, we need to prepare 10 specimens. Each specimen is loaded under a different fatigue load until a cyclically stabilized hysteresis loop is achieved. After obtaining ten cyclically stabilized hysteresis loops from the ten specimens, a cyclic stress and strain

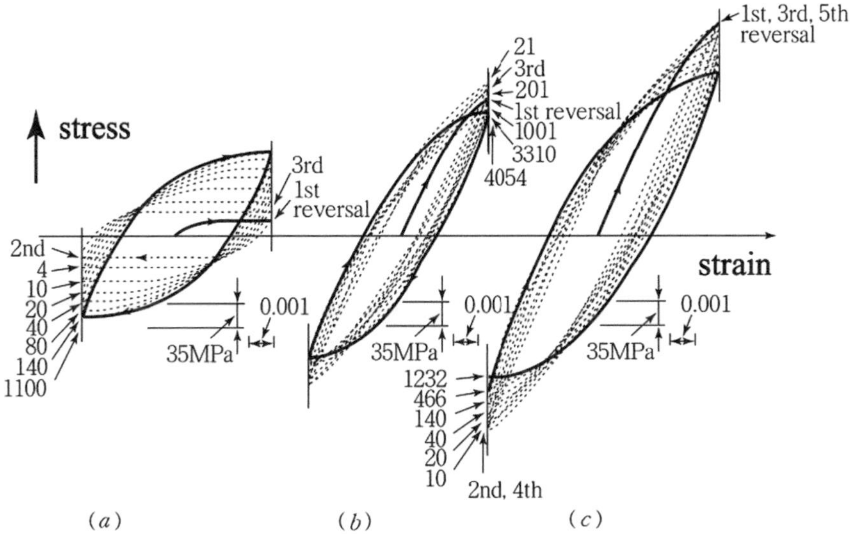


**Fig. 8.9** Cyclically stabilized hysteresis loop: **a** cyclic strain hardening, **b** change from cyclic strain hardening to cyclic strain softening, and **c** cyclic strain softening (Reprinted with permission from J. Morrow, "*Cyclic Plastic Strain Energy and Fatigue of Metals,*" in Internal Friction, Damping, and Cyclic Plasticity, ASTM STP 378, American Society for Testing and Materials, Philadelphia, p. 45, 1965, copyright ASTM International, 100 Barr Harbor Drive, West Conshohocken, PA 19428)

curve is constructed by connecting the vertices of the cyclically stabilized hysteresis loops, and then connecting it to a monotonic stress and strain curve, as shown in Fig. 8.10.

To adopt the single-specimen method, you must have a programmable dynamic UTM (Universal Testing Machine) that can apply a programmed fatigue load. To obtain the cyclic stress and strain curve with the programmed fatigue load, you should program the load in the following way,

1. The cycle load should have zero mean stress.
2. The cycle load should start from zero, increase gradually, then decrease gradually, and finally return to zero. This creates a block of the programmed fatigue load. This block load is applied continuously until the hysteresis loop no longer changes.

The testing result under the programmed fatigue load, as described above, is shown in Fig. 8.11. The ridgeline represents the cyclic stress and strain curve, which is very similar to the one obtained from the multi-specimen method.

***Example 8.4*** Why should the two cyclic stress and strain curves obtained from the multi-specimen method and single-specimen method be very similar?

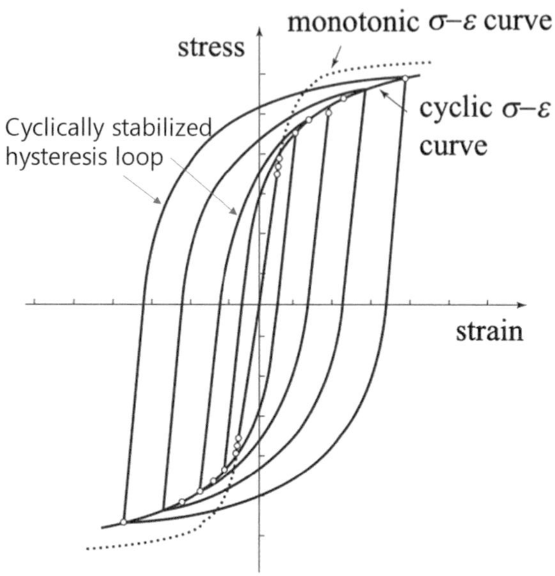


**Fig. 8.10** Cyclic stress and strain curve obtained from multi-specimens (Reprinted with permission from R.W. Landgraf, *Achievement of High Fatigue Resistance in Metals and Alloys,* ASTM STP 467, 1970, p. 3, copyright ASTM International, 100 Barr Harbor Drive, West Conshohocken, PA 19428)

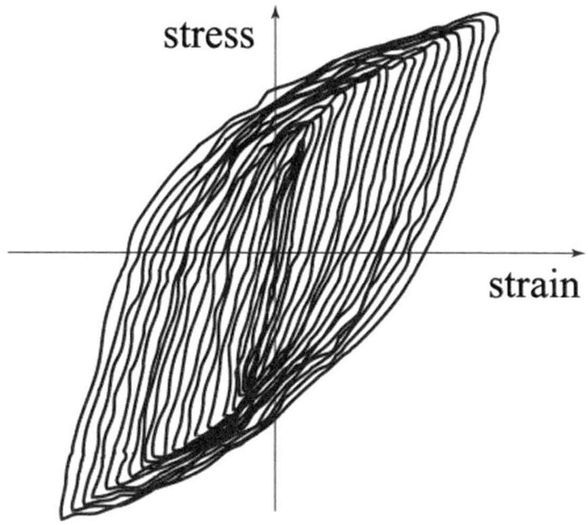


**Fig. 8.11** Cyclic stress and strain curve

**Solution**

To understand the reason, you need to recognize that countless different constant fatigue loads can be obtained from the programmed fatigue load mentioned above.

It is noted that Fig. 8.12a shows cyclic strain hardening, while Fig. 8.12c shows cyclic strain softening. However, Fig. 8.12b shows cyclic strain softening under low constant strain fatigue load but cyclic strain hardening under high constant strain fatigue load.

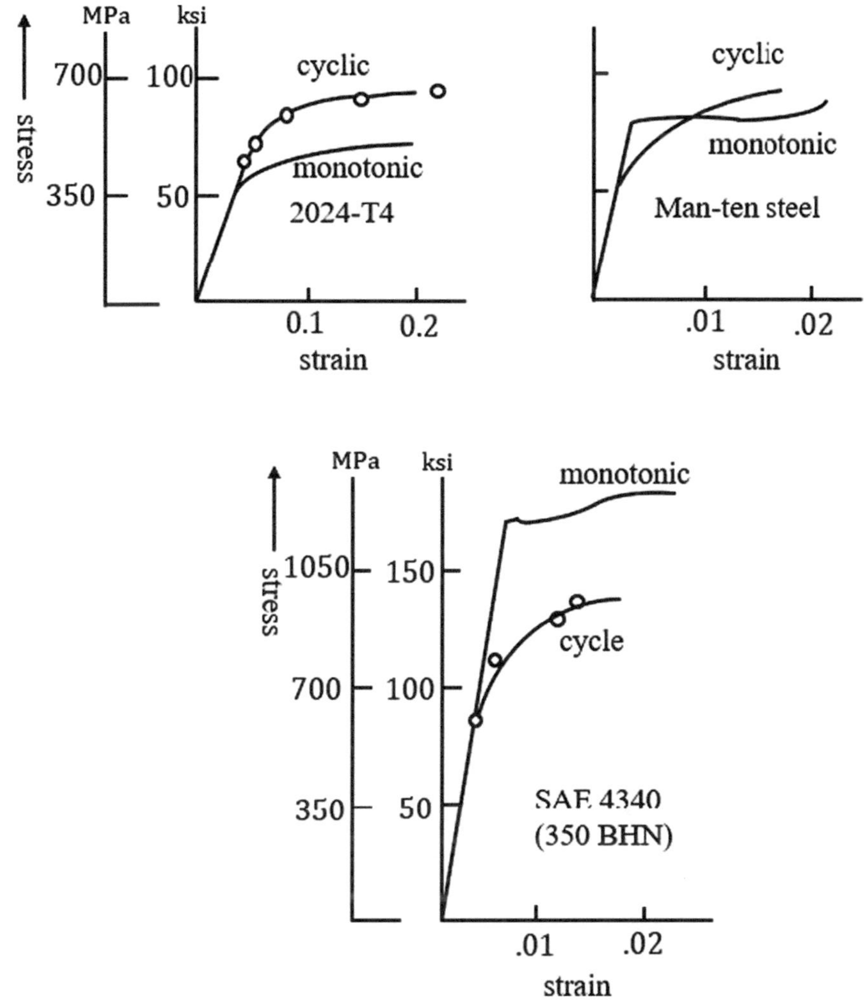


**Fig. 8.12** Comparison between monotonic and cyclic stress and strain curves (Reprinted with permission from Wundt, B.M., *Effects of Notches on Low Cycle Fatigue*, STP-490, American Society for Testing and Materials, Philadelphia, 1972)

Power law hardening in monotonic stress and strain curve is expressed as follows,

$$\sigma = K\varepsilon_p^n \tag{8.2}$$

where $\sigma_{ys} < \sigma < \sigma_u$.

$\sigma_u$ is the ultimate stress, $\varepsilon_p$ is the plastic strain, $K$ is the strain hardening coefficient, and $n$ is the strain hardening exponent. Strain hardening coefficient and strain

**Table 8.1** Monotonic and cyclic properties of materials

| Material | Condition | $\sigma_{ys}/\sigma_{ys}'$ (ksi) | $n/n'$ | $\varepsilon_f/\varepsilon_f'$ | $\sigma_f/\sigma_f'$ (ksi) |
|---|---|---|---|---|---|
| SAE 1015 | Normalized, 80 BHN | 33/35 | 0.26/0.22 | 1.14/0.95 | 105/120 |
| SAE 950X | As received, 150 BHN | 50/49 | 0.16/0.134 | 1.06/0.35 | 109/91 |
| SAE 1045 | Q + T (120°F), 225 BHN | 92/60 | 0.13/0.18 | 1.04/1.0 | 178/178 |
| SAE 1045 | Q + T (700°F), 410 BHN | 198/120 | 0.076/0.146 | 0.72/0.60 | 270/270 |
| SAE 1045 | Q + T (360°F), 595 BHN | 270/250 | 0.071/0.13 | 0.52/0.07 | 395/395 |

hardening exponent are material constants that should be measured using a standard method. Similarly, power law hardening in the cyclic stress and strain curve is expressed as follows,

$$\frac{\Delta\sigma'}{2} = K'\left(\frac{\Delta\varepsilon_p}{2}\right)^{n'} \tag{8.3}$$

where $\Delta\sigma'$ is the stress range, $\Delta\varepsilon_p$ is the plastic strain range, $K'$ is the cyclic strain hardening coefficient, and $n'$ is the cyclic strain hardening exponent. Cyclic strain hardening coefficient and cyclic strain hardening exponent are also material constants that should be measured using a standard method.

Table 8.1 shows the monotonic and cyclic properties of some materials.

In Table 8.1, $\sigma_{ys}$ and $\sigma_{ys}\prime$ is the monotonic and cyclic yield stress, respectively. $\varepsilon_f$ and $\varepsilon_f\prime$ is the monotonic and cyclic failure strain, respectively. $\sigma_f$ and $\sigma_f\prime$ is the monotonic and cyclic failure stress, respectively.

The criteria to distinguish between cyclic strain hardening and softening materials are as follows,

1. If $n > 0.2$ or $\sigma_u/\sigma_{ys} > 1.4$, the material generally shows cyclic strain hardening.
2. If $n < 0.1$ or $\sigma_u/\sigma_{ys} < 1.2$ the material generally shows cyclic strain softening.

However, these criteria can only be applied to some metallic materials, but it is not usually applicable to nonmetallic materials.

## 8.3 Approximation of S–N Curve

The standard S–N curve presents the relationship between stress amplitude and failure cycle, which is obtained under $\sigma_m = 0$. The simple method to obtain the approximated S–N curve under $\sigma_m = 0$ for metallic materials is given below.

$$S_e = \begin{cases} 0.5\ S_u & (S_u \le 200\text{ ksi}) \\ 100\text{ ksi} & (S_u > 200\text{ ksi}) \end{cases}$$

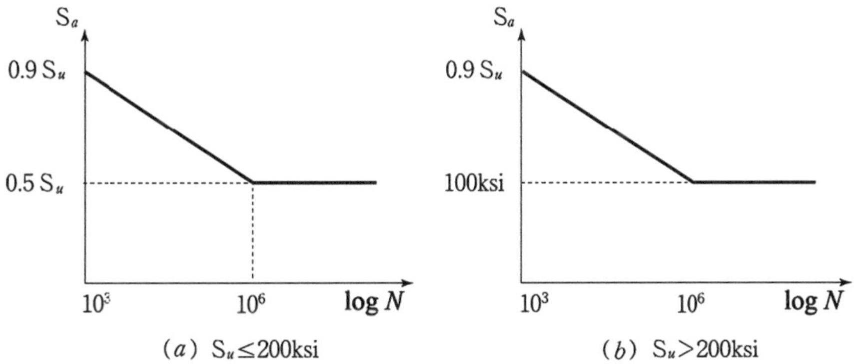


**Fig. 8.13** Approximation of S–N curve **a** $S_u \leq 200$ ksi, **b** $S_u > 200$ ksi

$$S_{1000} = 0.9\ S_u \tag{8.4}$$

$$S_u(\text{ksi}) = 0.5\ \text{BHN}$$

where $S_e$ is the fatigue limit, $S_u$ is the ultimate strength, $S_{1000}$ is the stress amplitude at failure cycle $10^3$, and BHN is the Brinell hardness number.

The approximated S–N curves obtained by Eq. 8.4 are shown in Fig. 8.13.

Since this approximation of the S–N curve saves a lot of testing time, it is a good method when no measured S–N curve is available. It is noted that this method requires only tensile testing or hardness testing, which is very simple.

## 8.4 Effect of Mean Stress

Figure 8.14 is a Haigh diagram, which shows the relationship among stress amplitude, mean stress, and failure cycles. It demonstrates that there are many fatigue loads consisting of stress amplitude and mean stress that result in the same fatigue life.

There are the following four types of Haigh diagrams for fatigue cycles, $10^6$,

(a) Soderberg $\dfrac{\sigma_a}{S_e} + \dfrac{\sigma_m}{S_{ys}} = 1$

(b) Goodman $\dfrac{\sigma_a}{S_e} + \dfrac{\sigma_m}{S_u} = 1$

(c) Gerber $\dfrac{\sigma_a}{S_e} + \left(\dfrac{\sigma_m}{S_u}\right)^2 = 1$

(d) Morrow $\dfrac{\sigma_a}{S_e} + \dfrac{\sigma_m}{\sigma_f} = 1$ (8.5)

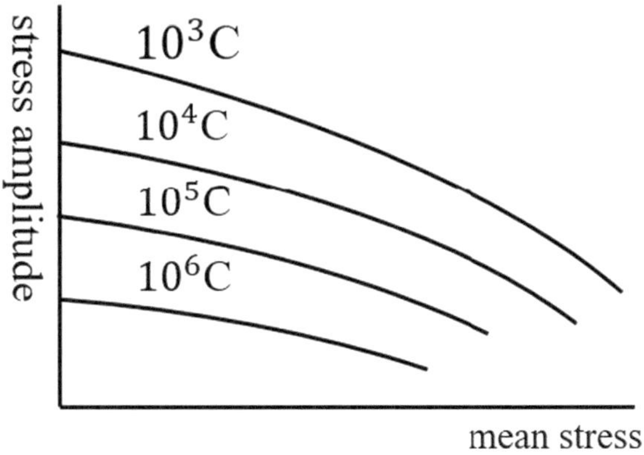


**Fig. 8.14** Haigh diagram

It is noted that in Fig. 8.15, the mean stress $\sigma_m$ can be either compressive or tensile. Equation 8.5 is obtained from the standard specimen.

Experimental data are usually located between the Goodman and Gerber diagrams. However, Goodman diagram is most popular because of simplicity.

It is noted that no fatigue failure occurs under the fatigue load in the region below the curve, while finite fatigue life is obtained under the fatigue load in the region above the curve. Haigh diagram equations corresponding to arbitrary fatigue cycle $N$ can be obtained by replacing $S_e$ with $S_n$ in Eq. 8.5. $S_n$ is the stress amplitude that gives an arbitrary failure cycle $N$ under zero mean stress.

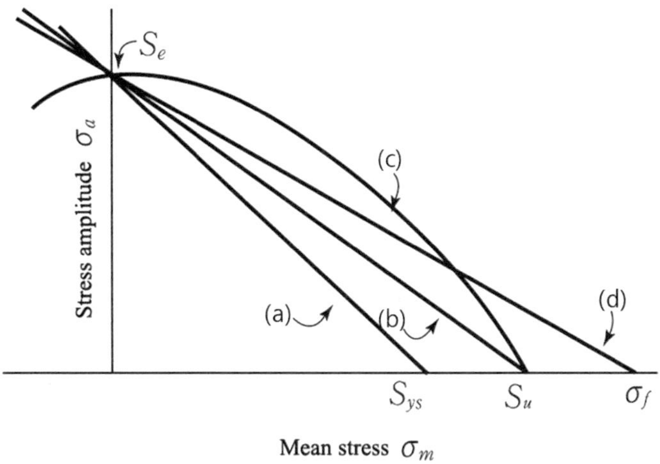


**Fig. 8.15** Four types of Haigh diagram at fatigue life, $N = 10^6$, **a** Soderberg, **b** Goodman, **c** Gerber, **d** Morrow

## 8.5 Random Fatigue Load

We can show that a random fatigue load consists of several constant fatigue loads. Therefore, we have dealt with constant fatigue loads up until now.

The rainflow cycle counting method is the most commonly used method to obtain these constant fatigue loads from a random fatigue load.

Usually, the random fatigue load is not completely random from start to finish, but rather random for a certain period of time, with the randomness continuously repeated. Therefore, you only need to focus on the random fatigue load applied for a specific period, which is one block in the full random fatigue load.

If you draw a hysteresis loop from one block of the random fatigue load, you will find that it is formed by hysteresis loops of various different sizes.

The same results can be obtained by the rainflow cycle counting method. This method is not explained in this book.

To estimate the fatigue life, Miner's rule is applied. It is expressed as follows,

$$\sum \frac{n_i}{N_i} = 1 \quad \text{Miner's rule} \tag{8.6}$$

where $n_i$ is the applied cycle number and $N_i$ is the failure cycle number under constant fatigue load.

It is noted that in Table 8.2, constant fatigue load $S_i$ can be fatigue loads with arbitrary $\sigma_m$. These constant fatigue loads $S_i$ can be obtained from the hysteresis loops obtained by using the rainflow cycle method. The failure cycle number $N_i$ under the constant fatigue load $S_i$ can be obtained by using Goodman diagram.

Equation 8.6 does not show the dependence of the order of applying constant fatigue loads, but it should be dependent.

In case of the high-low fatigue load test applied from high to low fatigue loads in order,

$$0.5 < \sum \frac{n_i}{N_i} < 1 \tag{8.7}$$

In case of the low–high fatigue load test applied from low to high fatigue loads in order,

**Table 8.2** Miner rule

| Constant Fatigue load | Applied cycle number | Failure cycle number |
|---|---|---|
| $S_1$ | $n_1$ | $N_1$ |
| $S_2$ | $n_2$ | $N_2$ |
| $S_3$ | $n_3$ | $N_3$ |
| $\vdots$ | $\vdots$ | $\vdots$ |

$$1 < \sum \frac{n_i}{N_i} < 2 \tag{8.8}$$

Equations 8.7 and 8.8 mean that Miner's rule should depend on the sequence of the applied fatigue loads. Therefore, many researchers have attempted to include the effect of the order of fatigue loads in Miner's rule and have suggested various modified versions of Miner's rule. However, Miner's rule is still considered the best due to its simplicity.

The following are the steps to estimate fatigue life under random load,

1. Obtain the constant fatigue loads by the rainflow cycle counting method from one block of random fatigue load.
2. Count the cycle numbers,$n_i$ corresponding to the constant fatigue loads, $S_i$.
3. Get $N_i$ corresponding to each constant fatigue load,$S_i$.
4. Count the number of repeated blocks.
5. Multiply that number obtained at step 4 by the cycle numbers, $n_i$ obtained in step 2 and replace the result with $n_i$.
6. Introduce $n_i$ and $N_i$ into Eq. 8.6 and compare the result with 1.

If the result shows that it is greater than 1, it means the material has already failed.

If the result shows that it is less than 1, it means that the material has not failed yet.

In case of less than 1, the remained fatigue life can be obtained by comparing the number of the block that gives equality in Miner's rule with the number of the block already used.

# Bibliography

## *For Part I Elasticity*


1. Timoshenko, S. P., & Goodier, J. N. (1951). *Theory of elasticity*. McGraw-Hill Inc.
2. Ugural, A. C., & Fenster, S. K. (1978). *Advanced strength and applied elasticity*. Elsevier North Holland Publishing Company Inc.
3. Chou, P. C., & Pagano, N. J. (1992). *Elasticity-tensor, dyadic, and engineering approaches*. Dover Publications Inc.


## *For Part II Fracture*


1. Hertzberg, R. W., Vinci, R. P., & Hertzberg, J. L. (2013). *Deformation and fracture mechanics of engineering materials* (5th ed.). Wiley.
2. Anderson, T. L. (2017). *Fracture mechanics-fundamentals and applications* (4th edn). CRC Press.
3. Murakami, Y. (Editor-in-chief) (1987). *Stress intensity factors handbook*. Pergamon Press.


## *For Part III Fatigue*


1. Bannantine, J. A., Comer, J. J., & Handrock, J. L. (1990). *Fundamentals of metal fatigue analysis*. Prentice-Hall Inc.





K. Y. Lee, *Introduction to Elasticity, Fracture and Fatigue*,
https://doi.org/10.1007/978-981-96-3600-6

# Index




K. Y. Lee, *Introduction to Elasticity, Fracture and Fatigue*,
https://doi.org/10.1007/978-981-96-3600-6

MIX
Papier aus verantwortungsvollen Quellen
Paper from responsible sources
FSC® C105338

If you have any concerns about our products,
you can contact us on
**ProductSafety@springernature.com**

In case Publisher is established outside the EU,
the EU authorized representative is:
**Springer Nature Customer Service Center GmbH**
**Europaplatz 3, 69115 Heidelberg, Germany**

Printed by Libri Plureos GmbH
in Hamburg, Germany